*The place of technology in social affairs is never neutral. But it is less neutral in some areas than in others. Writing with deep empathy and evocation for the ordinary people in history for which he has become so uniquely capable, Charles van Onselen tells the story of the role of the locomotive in regimenting, deceiving, ensnaring, holding, destroying, indeed sucking in and puffing out, the thousands of Mozambican miners who came to work the mines of South Africa in the early 20th century. Nelson Mandela named his presidential residence in Pretoria Mahlamba Ndlopfu (Tsonga for 'new dawn') in honour of the people of southern Mozambique who made (some in) South Africa prosper. Charles van Onselen documents why.*

WILMOT JAMES, VISITING PROFESSOR AT COLUMBIA UNIVERSITY

*Occasionally, social history research shines a piercing light on the entanglement of transport and society. Van Onselen's dazzling study of just one train route is about journeys loaded with fear, loathing and contempt. The Night Trains is a devastating account of human burden and wreckage.*

GORDON PIRIE, AFRICAN CENTRE FOR CITIES, UNIVERSITY OF CAPE TOWN

# The Night Trains

Moving Mozambican Miners To and From South Africa,
circa 1902–1955

Jonathan Ball Publishers

JOHANNESBURG & CAPE TOWN

© Text Charles van Onselen, 2019
© Published edition Jonathan Ball Publishers, 2019

Published in South Africa in 2019 by
JONATHAN BALL PUBLISHERS
A division of Media24 (Pty) Ltd
PO Box 33977
Jeppestown
2043

This limited edition reprinted in 2020, 2021 and 2023

ISBN 978-1-86842-994-3
ebook ISBN 978-1-86842-995-0

*Every effort has been made to trace the copyright holders and to obtain their
permission for the use of copyright material. The publishers apologise for any
errors or omissions and would be grateful to be notified of any corrections that
should be incorporated in future editions of this book.*

Twitter: www.twitter.com/JonathanBallPub
Facebook: www.facebook.com/JonathanBallPublishers
Blog: http://jonathanball.bookslive.co.za/

Cover by Michiel Botha
Design and typesetting by Triple M Design
Set in 11,5/17 pt Adobe Garamond

Printed by CTP Printers, Cape Town

*For the children of the Sul do Save,*
*a tale about their forebears*

# Contents

# Introduction

*Everyone who works on trains knows they have personalities,*
*they're like people. They have their own mysteries.*
— SAM STARBUCK, *THE DEAD ISLE*

If forced to choose the three technological innovations of the early 19th century that had the most profound and lingering influence on the process of globalisation and the making of world history well into the first half of the 20th century, your list might include the telegraph, the locomotive and the steamship. And if you were pressed to couple your choices to three of the most important transformative economic, social and political issues that characterised the period 1850–1950, you might get away with saying the Industrial Revolution, imperialism and colonialism. But, then again, you might not – and for good reason. Some will want to know about the development of the capitalist state, the expansion of the franchise and a new democratic order or, the converse, about the Russian Revolution and the two world wars. That, too, would not settle the matter since all it really does is to suggest many more complex cross-cutting connections that are all worthy of the most serious examination.

Managing complexity at this level of abstraction is a difficult if not impossible task, and the search for such sweeping answers is probably best left to those of a more theoretical bent. A historian attempting to make sense of how so intimidating a set of variables plays itself out over a century is probably better advised to rein in his or her ambition, perhaps pick just one technical marvel of the great age of material progress, and then turn it into the axis around which many of the other variables can be spun, in order to create the illusion of a few 'explanations' that are not entirely unconvincing or lacking in evidence.

Making the steam locomotive, the railway and the story of one particular train journey the focal point of a short study in order to create such an illusion – and all history-writing is, in part, illusion-creating in that the historian seeks to conjure up an array of the human senses and place them in times past in a plausible way – may be a useful start. And the African continent might not be the worst place in which to set it. It was, after all, Cecil Rhodes, a man who saw himself at the centre of British expansionism, who linked imperialism and the railway in a memorable phrase now totally eclipsed by nationalism: 'from Cape to Cairo'.

Rhodes' dream was of a railway stretching from the Cape to Cairo – a south–north dynamic that, we might note in passing, reveals how he mistakenly believed that a 'civilising' impulse and primary industrialisation might somehow be made to march in lockstep. It simply never happened and a single thoughtful glance at the De Beers' mine compound, in Kimberley, should have been enough to enlighten him. There is, however, little point in chasing his original idea via a factual inquiry, although it still has the makings of some good fiction. But might there not be some return from an investment in the historical imagination if we abandoned Rhodes' imperial dream of a south–north line running the length of continent, and focused instead on another railway, an east–west one, in southern Africa, driven in part by a nationalist vision that, ironically, drew much of its own political and ideological strength precisely from being opposed to Rhodes' aspirations?

It is significant that the idea of building a railway linking the South African Republic to Delagoa Bay, in Mozambique – Portuguese East Africa – was first floated in the 1870s, that is, after the discovery of diamonds in the northeastern Cape in the late 1860s had begun to transform economic prospects in the territory north of the Vaal River. President TF Burgers and his Boer notables, however, deeply suspicious of the potentially disruptive combination of British imperial ambitions and mining capitalism as factors underwriting expansion northwards, never saw their own rail link between Pretoria and Lourenço Marques (today Maputo) as the means by which they might develop a rival urban industrial base to that emerging in the diamond fields to the southwest. Rather, the idea was that, besides helping to ensure an economic and military supply line to a coast fairly free of British influence, the rail link might supplement the gradual, more ordered development of an agrarian republic that would benefit indirectly from the urban markets developing in the northeastern Cape.

But a weak concoction of maize and Mausers did little to enthuse investors in metropolitan Europe, and the idea behind an east-coast line languished for more than a decade until the magical third ingredient – small quantities of gold – was added to the mixture to make it more potent. The new mixture became explosive, however, as soon as the world's largest goldfields were discovered on the Witwatersrand in 1886. The force of the blast flung the original concept, upside down, into the surrounding countryside, and instead of the link – which was opened in 1895 – serving a sluggish agrarian economy, its *raison d'être* was transformed into serving a mining industry that gave birth to a new, expanding, urban-industrial dispensation.

3

This revolutionary change in the underlying purpose of the railway, from agrarian-industrial to industrial-agrarian, was effectively guaranteed and taken a step further after the British triumph in the Anglo-Boer War of 1899–1902. With the Rand mines so desperate for huge quantities of ultra-cheap labour that the Chamber of Mines eventually imported indentured labour from China, and the Portuguese at least as desperate to ensure rail traffic sufficient to underwrite the future of the port of Lourenço Marques, the colonising principals struck a series of deals at the expense of a third party – Africans drawn from south of the Save River (the Sul do Save) in Mozambique. In return for a guaranteed percentage of the rail traffic bound for the new goldfields, the Portuguese eventually granted the Witwatersrand Native Labour Association (WNLA) sole recruiting rights for cheap black mine labour in the Sul do Save, with the majority of miners to be transported to and from the Rand by WNLA trains.

But, regardless of whether the Eastern Main Line was to benefit an emerging agro-industrial or industrial-agrarian order, the one thing that the Johannesburg–Lourenço Marques railway was never intended for was the transportation of passengers. It was, and remained, pre-eminently a conduit for the movement of bunker coal from the Eastern Transvaal collieries down to the coast, and heavy mining equipment and machinery up to the gold mines on the southern Highveld. Yet move passengers it did – and on an industrial scale. It is estimated that between 1910 and 1960 some 5 million passengers were ferried between Booysens station in Johannesburg and the Komatipoort/Ressano Garcia station complex on the Mozambique–South Africa border. It was an extraordinary logistical feat but one that, rather strangely, has attracted very little interest from most labour historians, let alone anthropologists, economists or sociologists.

4

This apparent lack of curiosity is puzzling because, between them, the Eastern Main Line and the seemingly endless supply of black labour that it conveyed across the face of the southern African plateau formed the umbilical cord and lifeblood that gave birth to the mining revolution that took place on the Witwatersrand between the two world wars. Indeed, the easy and guaranteed access to cheap Mozambican labour – for close on half a century – contributed centrally to the profitability of the gold-mining industry that, in turn, drove South African economic prosperity and allowed its white ruling classes to indulge in moral and political electoral self-indulgence for more than a century.

Moreover, the logistical system that underpinned this mass, transnational sell-off of Africans into industrial servitude in exchange for rail traffic – an exchange that did very little for the long-term benefit of the people themselves – grew out of an unusual partnership on the South African side of the border that is equally intriguing, yet one that has also remained almost entirely unexplored. The WNLA, a company offshoot of the Chamber of Mines, was allowed to run a privately operated train on publicly owned tracks managed by the South African Railways (SAR) and its predecessors. This private-public partnership gave rise to a set of intermanagerial conflicts and tensions that were never fully resolved.

A private train running along public tracks effectively rendered the Eastern Main Line part of a legal no-man's land – part of a conveyor belt that was seldom inspected, let alone systematically so, over the *longue durée* by the state's health, safety or labour agencies. Nor, for that matter, was it directly policed by the South African Railway Police (SARP), who confined their law-enforcement duties largely to station precincts. Indeed, had WNLA trains and their passengers fallen directly under state scrutiny for a sustained period, those agencies might have picked up that the migrant workers were not only confined to carriages for a day and a half, undernourished and without easy access to potable water, but also lacked the most basic ablution facilities. More seriously still, in order to prevent the desertion of some migrants who had been fed into the system via forced-labour practices in Mozambique, the misleadingly named 'recruits' on the journey up from Ressano Garcia were frequently locked into railway carriages by WNLA conductors.

This neglect by the state of the condition of the private trains or the welfare of the migrant workers they carried – an indifference that extended fully to the SAR administration – did nothing to encourage the WNLA management to improve the 'service' that it ran. But, as the WNLA management discovered, to its cost, there was flip side to such insouciance. The SAR and its operatives, focused on moving freight through the system, were unwilling to accord a private train carrying black migrants the status of a passenger service entitled to the same priorities and privileges enjoyed by the regular mail train ferrying whites up and down the line. The resulting delays in delivering labour to the mines hampered industrial efficiency and proved to be a constant source of tension between SAR managers and the WNLA.

Differences in managerial ideologies, operational priorities and the terminology employed by the SAR and the Chamber of Mines when it came to categorising or referring to black workers were, however, never confined to timetables. In general, it was true that the railways were always willing to slow the flow of migrant traffic through the system, and more especially so as the SAR's catering arm focused on selling food and drink to cash-flush miners at stations on the route back to Ressano Garcia. But WNLA managers, usually striving for industrial efficiency in their delivery of men to the mines, always wanted to increase the tempo and improve the turnaround times for their trains.

The great promise of the steam locomotive was that it provided the power and speed needed to conquer distance. But, when it came to the functioning of the long-distance labour trains running between Johannesburg and Ressano Garcia, there were contradictory forces at work that stemmed from racist thinking. The SAR was inclined to view the black migrant trains as carrying goods, or 'human freight', and did little to expedite their movement through the system. The WNLA managers shared much of this thinking but, for reasons of industrial efficiency, wanted their trains to be treated as passenger trains. The net result was that the supposed advantage offered by the steam locomotive – its capacity to conquer distance through speed – was severely compromised.

The disagreement over whether migrant workers in transit should be treated or seen as mere commodities – inanimate things – or as human beings fully deserving of the status of passengers, had consequences that went well beyond timetabling issues. It was the official designation of a train as 'freight', 'passenger' or 'mixed' that determined the level of competence, experience and skill required of a locomotive driver. The driver's understanding of the nature of the train he was operating – black or white, goods or freight, or mixed – thus also determined the speed at which he operated the locomotive. All of this held the potential for confusion and uncertainty, which in turn was capable of compromising the safety of the train and its passengers. In short, in a society steeped in racial thinking there was no easy way to seal off the operational realities of the railway system from the broader patterns of racial prejudice to be found in the society at large. These southern African peculiarities further complicated any understanding of the cause and nature of accidents, including fatal accidents, involving the labour trains.

None of these considerations, however, provide the historian with a ready-made template for recording the history of a colonial labour train or, to be more precise, for the writing of a history of a set of inward- and outward-bound train journeys as they might have been experienced by black miners who were, for the most part, illiterate. Nor, for that matter, are there existing models elsewhere that suggest how best to go about trying to conjure up the experience of the return, long-distance, transnational journeys of migrant workers without relying on first-hand oral testimony instead of the surviving – incomplete – documentation of their employers. Any such exercise is bound to be deeply flawed and, at best, might be considered to be an experiment – and more likely a failed experiment.

And yet it has to be attempted, because sometimes the emerging consciousness of 'national' history is shaped at least partly by stories conveniently forgotten, or tales deliberately suppressed, in an attempt to avoid awkward, painful narratives that cannot readily be reconciled with banal, ideologically driven notions of linear 'progress'. The motto 'Ever Forward! Ever Upward!' may be inspirational for untutored minds but it has almost nothing to contribute in half-decent schools of history. In a chapter in a popular history of the Witwatersrand gold mines published in the glorious year of the coming of the South African republic, 1961, or in the subsequent editions, there is simply no mention of the WNLA trains linking Ressano Garcia and Johannesburg, nor of the predominance of Mozambican labour.[1]

It would be a serious mistake not to try and record, no matter how imperfectly, a set of experiences that go to the heart of the Mozambican contribution to a South African industrial revolution that stalled and then failed completely to incorporate or extend its skewed benefits to the majority of the population. The advent of the railway in the New World, and the success that flowed from its having 'opened up' not only the United States and Canada but many other countries, is well-known and well-celebrated. Yet we know, too, from bitter historical experiences in Germany, Poland and Turkey, to name but a few, that, when employed

8

under different circumstances by those bent on conquest, evil, gross exploitation or self-enrichment, the steam locomotive and train are capable of underwriting genocide, human degradation and the oppression of the vulnerable. Indeed, there seems to be something in the very design and operation of the railway, the steam locomotive and the train as a form of temporal mobile incarceration that lend themselves to geopolitical thinking, social engineering and marginalising the vulnerable.

Mozambicans moving in and out of South Africa were never the object of a genocidal impulse on the part of those who controlled the host country economically, socially or politically. The men of the Sul do Save were, however, first pushed and then pulled into the front line of an industrial war of staggering proportions for cheap black labour in the coal- and gold-mining centres. Moreover, there was something about the workers' experiences on both the inbound and outbound trains that suggested a potential for greater evil when mass transit systems were employed solely for profit in a colonial setting scarred by overtly racist thinking.

Besides being sealed for much of the journey from Ressano Garcia to Booysens – a telling indictment of what allegedly was an entirely voluntary labour supply – the inward-bound WNLA trains were, until well into the 1920s, capable of inflicting serious bodily harm on migrant workers. Using boxcars, cattle trucks and coal trucks that lacked the most basic facilities could turn the journey west into a full-scale ordeal for miners until the gradual introduction of third-class carriages after the advent of Union, in 1910. But, even then, the modern coaches were occasionally supplemented by bogies designed for the conveyance of animals or freight rather than people right up to the Great Depression (1929–1933).

Hungry, clothed like Tressell's *Ragged-Trousered Philanthropists* and perfectly penniless, the 'recruits' on the 804 up-train attracted almost no outside interest as they passed slowly through the various stations west.

On the Rand, the Chamber of Mines was aware that it was engaged in traffic that some might consider shameful, and that it could become the subject of unwanted criticism. European voters were content to benefit from a racially oppressive mining system but did not want to have to contend with the visual pollution that arose when they were forced to acknowledge the presence of cheap black labour making its way through the streets of supposedly all-white cities. The success of the system that, by the late 1950s, saw over 400 000 rurally based workers moving in and out of the Witwatersrand every year lay, in good part, in keeping it as hidden as best possible. And, for the most part, it worked. As one historian rightly noted in 1961, the system works 'so smoothly' that 'the average [white] South African does not even know that these multitudes move in and out of Johannesburg every year as 'time-expired' men [and] are replaced by newcomers'.[2] Whenever cost-effective and practically feasible, as regarded station layout and timetabling, the SAR and the WNLA cooperated to ensure that the processes of entraining or detraining black labour in Johannesburg and Lourenço Marques was done as discreetly as possible.

Generally, the WNLA trains ran under cover of darkness, with the lesser part of the journeys, by far, falling within daylight hours. This was largely for logistical reasons because the busy Eastern Main Line, like many others, was marginally easier to negotiate at night, when there was less traffic than during the day. The disadvantage was that the most treacherous part of the route – the mid-part of a modern Middle Passage – from Waterval Boven down to Waterval Onder had to be negotiated at midnight. The train accidents that gave rise to the largest number of fatalities among returning workers, those in 1918 and 1949, occurred in the small hours of the morning. Accidents taking place in the middle of the night did nothing to facilitate search and rescue operations.

But having a train, replete with 'hospital coaches' – served neither by doctors nor by nurses – slithering through the Highveld darkness held other, indirect benefits for those operating the scheme. Not only was that which was out of sight also largely out of mind in a political dispensation that enjoyed the unending and overwhelming support of a socially blind white electorate, but also those operating an oppressive system on a day-to-day basis – the WNLA train conductors – found it easier to manage the discontent and chronic 'restlessness of the natives' while the migrants' circadian clocks were compromised. Disoriented, sleepy workers were less likely to object to, or resist, a dehumanising mass-transit experience.

The underground industrial war waged on the Witwatersrand was at its most intense between 1902 and the mid-1920s, and silicosis and tuberculosis claimed the lives of tens of thousands of black Mozambicans crammed into male-only mine compounds. In appearance and function alike, the compounds were more akin to prisons than they ever were to conventional 'single-quarters' working-class housing, and the social and sexual behaviours that they facilitated likewise resembled those usually associated with prolonged periods of incarceration. The physical and mental toll that this inflicted on workers was reflected in both the design and organisation of the 'hospital coaches', which had dedicated 'wards' for the diseased and the disabled and a special sealed compartment for the insane.

For the many relatively cash-flush, time-expired workers on the train back to Ressano Garcia each Thursday, it was a joyous occasion that heralded the chance to be reunited with family and loved ones in the Sul do Save. But, for scores, nay, hundreds of others the train journey was more like a mass evacuation of the living dead and the walking wounded from a war zone. For many years before and long after the First World War, it was not unheard of for a black miner to have been 'found dead' at various points on the 307 down-train as it moved slowly eastward through

Waterval Boven, Nelspruit and Kaapmuiden. Not always easily identifiable – other than by a metal disc worn around the neck – corpses were removed from the train along the line or at Komatipoort and handed over to the police. From there they were passed on to the district surgeon for post-mortem examination, during which the diseased lungs might be removed and sent back to Johannesburg for medical research. What remained was buried in the local, racially segregated cemetery.

But even for those healthy Mozambicans carrying home their modest, boxed personal possessions and cash savings or only the most recent wages – or the same on behalf of brothers still bound by contracts that were criminally enforceable – the down-journey back to Ressano Garcia necessitated running a gauntlet. At every station where the train stopped, and sometimes so for deliberately prolonged unauthorised stops at Germiston, Witbank, Waterval Boven, Nelspruit, Kaapmuiden and Komatipoort, there were men, women and children – black and white South Africans alike – eagerly wanting to profit from the passing trade. For confidence tricksters, currency-exchange crooks, hawkers, hucksters, robbers, stationmasters, thieves and traders peddling food and drink or trinkets to the hungry and sleep-deprived, the 307 down represented little more than a biweekly bank-on-wheels.

The tens of thousands of men drawn from the Sul do Save, the indispensable filling in an economic sandwich put together for global, risk-averse investors by the Chamber of Mines on the Witwatersrand and successive Mozambican administrations, had few genuinely concerned friends in high office on either side of the border. In truth, they had not many more sympathisers among ordinary black and white South Africans, who were so busy sizing up each other's respective interests in the long, bad 20th century that they had little time for worrying about 'foreign natives'. Despite this, there was clearly something about the plight of those coming to and going from the mines by train – and not only from the Sul do Save – that still haunts some South Africans of all colours. The clearest indication of this lies in the poignant lyrics of '*Shosholoza*' and '*Stimela*' – two songs that an aspirant 'nation' has taken ownership of, even if those who now sing them have little understanding of 'the train'. And in that lies a great danger – of forgetting easily over remembering correctly.

Nationalist politicians in many countries, often as incoherent as they are illiterate, find playing with the past in public almost irresistible, because it allows for moral posturing about 'our' heritage without taking meaningful responsibility for safeguarding and studying history, which is often more authentic and enduring since it is largely a private activity. Heritage is often about appealing to the heart, about providing answers, doing drama and performance, and solidifying an electoral base; it is part of a narrowing and mobilising strategy. History, when done well, is an appeal to the mind, and is about debate, contingency and questioning received wisdoms in ways that deepen our appreciation and understanding of who we are and why and how we did certain things, and perhaps even allows us to learn.

Neither Mozambique nor South Africa now wants for heritage-peddlers, many of them wholly owned bureaucratic lackeys of nationalist politicians dispensing patronage and trading shamelessly in their own real, or more often imagined, versions of the past. Unfortunately, insofar as it relates to transnational migrant labour between the two countries, the mobilisers and the muddlers have now discovered one clear-cut, almost exclusively visual, element in the history of the train – the site of the 1949 disaster at what once was Waterval Boven (today Emgwenya). To the accompaniment of the usual hollow public performance and pronouncements, a commemorative stone was erected there in 2006 to remember those who lost their lives in the accident.[3] For those taken with nationalist mythology, the site and stone alike have all the makings of good 'heritage' even if they make for history incorrectly told and poorly remembered. The men, women and children of the Sul do Save deserve so much better. This little book is a story for them, lest we forget and lose forever the opportunity of remembering how the sad events that befell the train at Waterval Boven were illustrative of a system, not an 'accident'.

# PART I

# CHAPTER ONE

# The Eastern Main Line

## Umbilical Cord of South Africa's Mining Revolution circa 1902–1955

*Railroad trains are such magnificent objects we commonly mistake them for Destiny.*

— EB WHITE

If you insert yourself among the sprawling social ruins that connect the supposedly separate universes of town and countryside in southern Africa – and you listen carefully enough – you may catch the outlines of a poignant refrain or two. Mournful lyrics remind you why, after a stalled industrial revolution scarred by crude divisions of class and colour, we probably need to be reminded of how the very act of migration – that ceaseless rotation of bodies between their places of birth and the subterranean vortices of wealth that opened up among indigenous peoples after 1870 – was central to the plight of those who were pushed into giving most and gaining least from the coming of a new order built almost exclusively on the relentless maximisation of profit.

Elsewhere across the world, amid similar upheavals, contracting economic sinews drew men and women from preindustrial dispensations into broadly contiguous, regionally defined patterns of work and residence.

The first Industrial Revolution saw no great physical distance separating most farmers from their field hands, factory owners from workers, pit owners from miners, or even shipowners from ports. But, in colonial Africa, political power and racial difference were aligned to meet the demands of industries locked into financial markets spread around the Atlantic periphery and whose local agents contrived to keep rich and poor, white and black, in separate domains. Here, more so than in almost any comparable case in modern history, the spatial articulation of capital and labour was manipulated so as to benefit the emerging economic centre while simultaneously promoting the political and social separation of a privileged, enfranchised white minority from an unenfranchised black majority. The older, more familiar, centripetal economic dynamic was offset by centrifugal political forces that gave rise to an inherently antagonistic tendency that was as counterintuitive as it was counterproductive. Such an unusual configuration was possible only with the aid of the most modern technology. Steam-driven locomotives, puffing along well-defined pathways determined by colonial administrations, shuttled to and fro in a ceaseless attempt to minimise costly differences of time and space.

Nowhere was the physical hardship and psychological pain occasioned by the movement of migrant workers more evident than along the international line linking colonial Mozambique and South Africa. It was the enduring memory and trauma of rail journeys that commenced at the port city of Lourenço Marques, in Delagoa Bay, and proceeded through the roundly resented coalfields near the Eastern Transvaal town of Witbank (eMalahleni) before eventually terminating in the goldfields of Johannesburg (Egoli) that chiselled their way into the souls of black men and women from right across southern Africa. Around 1910, by which time the railway was already firmly embedded in African imaginations, one curious European was moved to record a fragment of what he called an 'epic poem' recited by his servant. The train was, the young black man chanted, 'the one that roars in the distance ... the one that crushes to pieces the warriors and smashes them!'[1] Just two decades later, in KwaGuqa, the African township that serviced greater white Witbank and gave us one of the country's finest jazz musicians, men and women had long been singing a song destined to become famous throughout the troubled land.

The lyrics and rhythm of '*Shosholoza*' evoke such a powerful image of *stimela*, the steam-driven locomotive ferrying Mozambican and other workers back home through the mountains, that, rather unusually, the song has entered the repertoire of people of all colours right across the region, even though its deepest meaning is often lost on the well-heeled. Perhaps it was the very words of the song – '*Shosholoza, wen' uyabaleka, wen' uyabaleka, kuleza ntaba, stimela siphume South Africa*' – that subconsciously primed the great Hugh Masekela, once of Witbank, when, one night in 1973, during his exile in the United States, the tempo and lyrics for his own composition, '*Stimela*', tumbled forth from his mind and lips in what he later recalled as a single, near-unstoppable deluge:[2]

> There is a train that comes from Namibia and Malawi
> there is a train that comes from Zambia and Zimbabwe,
> There is a train that comes from Angola and Mozambique,
> From Lesotho, from Botswana, from Swaziland,
> From all the hinterland of Southern and Central Africa.
> This train carries young and old, African men
> Who are conscripted to come and work on contract
> In the golden mineral mines of Johannesburg
> And its surrounding metropolis, sixteen hours or more a day
> For almost no pay.

Deep, deep, deep down in the belly of the earth
When they are digging and drilling that shiny mighty evasive stone,
Or when they dish that mish mesh mush food
   into their iron plates with the iron shank.
Or when they sit in their stinking, funky, filthy,
Flea-ridden barracks and hostels.
They think about the loved ones they may never see again
Because they might have already been forcibly removed
From where they last left them
Or wantonly murdered in the dead of night
By roving, marauding gangs of no particular origin,
We are told.
They think about their lands, their herds
That were taken away from them
With the gun, and the bomb, and the teargas, the Gatling and
   the cannon.
And when they hear that Choo-Choo train
A-chugging, and a pumping, and a smoking, and a pushing,
   a pumping, a crying and a steaming and a chugging and
   a whooo whooo!
They always cuss, and they curse the coal train,
The coal train that brought them to Johannesburg. Whooo whooo!

Stimela
Sihamba ngamalahle
Sivel' eDalakubayi
Sangilahla kwaGuqa
Bathi sizo-mba-malahle
sizo-mba-malahle

Iyohhh …
(Stimela)
Sidl' inyu enkomponi.
(Stimela!)
Sihleli njengezinja, siyelele mame
Emigodini, babe
Sikhalel' izihlobo zethu
(Masibuyele le! eDalakubayi)
Sikhalel' izingane zethu wololo!
(Masibuyele le! eDalakubayi)
Sikhalel' abazali bethu!
(Masibuyele le! eDalakubayi)
Sikhalel' abafazi bethu, sithi
Yelele yelele yelele yelele yelele
(Masibuyele le! eDalakubayi)

Stimela
Sihamba ngamalahle,
Sivel' eDalakubayi.

Helele bathi Stimela mawo
Stimela
Sihamba ngamalahle,
Sivel' eDalakubayi.

Stimela!
Sihamba ngamalahle,
Sivel' eDalakubayi.

There are now several versions of this haunting evocation of the incoming train and the fate of young men deposited in the most merciless of industrial settings. The one that perhaps gets closest to conjuring up the migrants' trauma, however, has lyrics that are rendered a cappella style. It is replete with telling phrases of how the coal-gobbling engine and open trucks coming from eDalakubayi – Lourenço Marques – simply dump their quota of unhappy migrants on the station platform at Witbank, from where they are swept up into a nearby compound before being dispatched into the blackened underbelly of the collieries. It ends with a plea for the unhappy workers to be returned to Lourenço Marques so that their longing for absent wives, children and parents might be sated. Unlike 'Shosholoza' – a catchy tune but one with far less overtly emotive appeal – the heartbreaking poetry of 'Stimela' is perhaps just too painful for it to be accommodated easily in the 'South African' psyche. There are other reasons for this; it sometimes leaves ethnic nationalists of all colours, of whom there is no shortage at present, or historically, just a mite uncomfortable.

Although the lyrics of 'Stimela' resonate strongly with the experience of *all* incoming migrants on the mines, and there are oblique references that speak powerfully to the specific nature of racial oppression in the apartheid era, it is equally clear that the train's passengers are *not*, in the main, black South Africans. In fact, they are outsiders, Africans drawn in from beyond the country's borders, men hauled in from across the central and southern regions of the continent, pulled into the work that, arguably, demanded more sweat and tears from them than it did from any other domestic grouping in order to secure the development of the mining industry. And, of all the outsiders, none contributed more to the wealth of South Africa than did the faceless migrants hauled in from Mozambique, men extruded from the province south of the Save River – the Sul do Save – and its dominant port city, eDalakubayi.

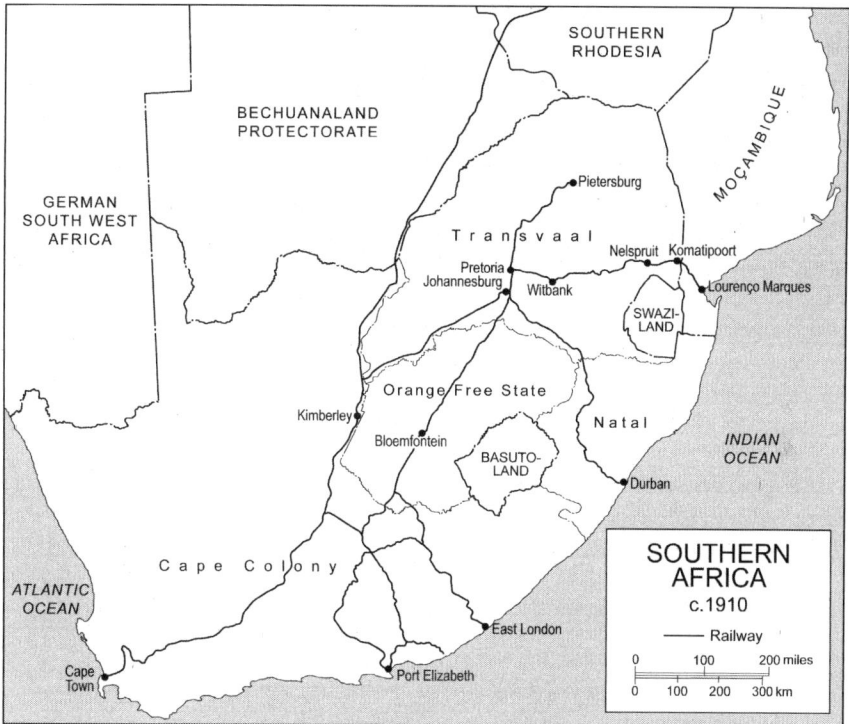

SOUTHERN
AFRICA
c.1910

It is tempting to see the Eastern Main Line connecting Johannesburg and Lourenço Marques as having been constructed exclusively, or even largely, so as to drag in the raw black muscle power of the politically powerless and the racially oppressed who had the misfortune to be caught between an industrialising South Africa and a compliant and venal Portuguese administration. Nevertheless, it would be misleading. If it was only cheap labour that the industrialists on the Highveld were after, the railroad would have been made to angle through Gaza, the heart of Thonga territory and the primary source of black workers, rather than being directed towards Delagoa Bay. The origins of the line, however, predated the largest mineral discoveries north of the Vaal River. The ambitious reasons that lay behind it, along with its slow appearance, are therefore deserving of some attention before we focus again on the train that ferried hearts and minds across the linked border outposts at Komatipoort and Ressano Garcia. There, two stations marked the spot where migrants crossed a latter-day River Styx separating the old and allegedly barbaric preindustrial ways of the black, the colonised and the weak from the world of the all-conquering, modernising, pseudo-paternalistic, white and strong.

CHAPTER TWO

# The Changing Economic Logic of the Eastern Main Line

## The Shift from an Agro-Industrial to an Industrial-Agrarian State

*No one is in charge of the process, this is what makes history so interesting,
it's a runaway freight train on a dark and stormy night.*

— TERENCE McKENNA

The discovery, in 1867, of diamonds in the vicinity of the Orange River, near Hopetown, prompted an economic awakening in the hitherto sleepy upper reaches of the Cape Colony. When yet more substantial gemstones were found just two years later the vigorous eye-rubbing that accompanied the escalating value of properties around Kimberley became a subject for debate even beyond the Vaal River. There, a few more ambitious, literate and market-attuned Boers felt that, under the guidance of a better-educated State President, the prospects of the landlocked South African Republic, still very much an agricultural backwater lacking markets, might perhaps be improved.

The choice of these progressive-minded Boers for the highest office in the land – questioned from the outset by a few conservative brethren notorious for having turned their backs on the influences of the Enlightenment and the Cape when slavery was abolished in the 1830s – fell upon a theolog-ically liberal and modernising candidate who appeared to be unelectable. Thomas François Burgers, educated at the University of Utrecht, held such philosophically advanced views that he had been charged with heresy and forfeited his position as a minister in the Cape Colony's Nederduitse Gereformeerde Kerk (Dutch Reformed Church) only to see the decision overturned on appeal, by the Privy Council, in 1867.[1]

In 1872, however, the Transvaal Boers were still largely unencumbered by the upheavals that would come after the discovery of gold in Lydenburg only months later and add fuel to Britain's smouldering imperial ambi-tions. An electorate, supposedly composed entirely of backwoodsmen hopelessly set in their Bible-bound ways, elected Burgers as president by an overwhelming majority. This allowed him, albeit briefly, to embark on a few enlightened educational projects that struggled to take root in an agrarian economy. It was the first and last time that republicans north of the Vaal took in a puff of 'liberalism', which never extended as far as influencing race relations. After that, with the genies of gold and British greed out of the bottle, the Boers never again entrusted their political destiny to a president strongly committed to modernisation. Burgers' progressive ideas for a country he sensed as needing to stake a claim in the new industrial world, which was already belching smoke in Kimberley, became rapidly uncoupled from those of a poorly educated, farm-based constituency alarmed by the spectre of being overtaken by urban outsiders intent on ushering in open competition for the cheap labour of Africans in an unfamiliar new world.

Undaunted, but with his policies already rapidly running out of steam, Burgers – buoyed by gold discoveries – set off for Europe in 1875, bent on raising capital to build a railway that would run from Delagoa Bay to the Transvaal border. The fact that the Cape-based, empire-friendly shipping magnate, Donald Currie, was equally convinced that Lourenço Marques had a promising future if linked to the southern African plateau by rail, would hardly have swayed the wary Transvaal Boers.[2] And, when a military campaign against the BaPedi in the northeastern Transvaal also went awry, Burgers' fate was all but sealed. Some of the rails acquired in Holland from capital raised through debentures were shipped to far-off Mozambique, but most were just left to rust in the old Zeeland port of Flushing (Vlissingen). In 1878, the British, in the shape of Sir Theophilus Shepstone and a score of mounted policemen, occupied Pretoria, the tiny, largely demoralised capital of a nearly bankrupt republic.

The unwanted British presence was, however, short-lived. A Boer victory in the first War of Independence (1881) coincided with the discovery of yet more promising streaks of alluvial and reef gold around Barberton, thereby helping turn a minor triumph into what, over the longer term, would prove to be a Pyrrhic victory. The imperial fox had been put out of the henhouse, but the hole in the fencing had just got larger and the menace was ongoing. Sensing the growing threat circling the country, President Paul Kruger and his conservative political allies had new and more urgent needs for wanting a secure passage to the sea that would be reasonably independent of the steadily expanding British influence.

In 1884, the Kruger government awarded a concession to build the Transvaal section of the Delagoa Bay railway to a Dutch-German consortium, which, three years later, led to the flotation of the Nederlandsche Zuid-Afrikaansche Spoorweg-Maatschappij (NZASM).[3] At the same time, across the border, a Portuguese concession to extend the line from Lourenço Marques to Ressano Garcia was handed to Colonel Edward McMurdo, an American who proved to be more interested in speculative profits than railway construction. Indeed, so tardy was McMurdo in meeting the stipulations as to tangible progress in his contract that the

Mozambican administration tired of him and nationalised the railroad only 36 months later. Under the supervision of the new Caminho de Ferro de Lourenço Marques (CFLM), the line was successfully extended to the border at Ressano Garcia by 1890.[4]

But all this stop-go, grab-and-hope and push-on-regardless activity was not confined to the malaria-infested flats and swamps across which the track was being laid in the Mozambican section of the line, although the staggering loss of life it occasioned hardly facilitated construction.[5] The Kruger government, too, was forced to contend with unexpected problems. Beyond the usual financial shenanigans that accompanied Victorian railroad construction lay the greatest disruption ever experienced in the region – the discovery, in 1886, of the Witwatersrand goldfields. Almost every year thereafter tended to confirm emerging opinion that it was the largest and most continuous low-grade gold reef in the world. As early outcrop diggings gave way to reef mining, so it transformed the prospect of the initial, quasi-speculative concessions granted as hopeful adjuncts to the development of an agrarian-based economy into the potentially huge money-spinners of a more established industrial order.

As in the case of the CFLM – and for essentially the same reason – the NZASM was confronted with administrative and political delays, the problem of raising capital at competitive interest rates in a rapidly changing economic environment, and the need to assemble competent engineering and technical teams. The net result was that the Delagoa Bay railway, first envisioned back in 1875, became a functioning reality only once the line was officially opened on 1 January 1895.[6]

It was the spectacularly swift rise of the Rand gold-mining industry that breathed financial life into the Eastern Main Line and rendered it reasonably profitable by the late 1890s. From its inception, the primary purpose of the link to Delagoa Bay was to operationalise the importation

of heavy mining equipment and timber props from Australia and the
United States to the coal and gold mines stranded on the Highveld – a
distant, treeless, grassland plateau.[7] Gradually integrated into the rest of
the developing national rail system, the Eastern Main Line also facilitated
the regional transportation of everything from perishable foodstuffs destined for Lourenço Marques to export goods and mail bound for other
port cities across the world. But, whatever hard-wired economic virtues
the new system encompassed, it was never designed to cater for, let alone
give preference to, the comfort of its passengers.[8] It was meant to – and
did – enable the wholesale transportation of commodities but was almost
totally indifferent to the needs of people. That same bedrock logic did
much to inform the design, construction and operational realities of the
new railway system.

In constructing the coastal linkage, NZASM engineers were called
upon to deal with familiar problems of varying heights and distance.
Johannesburg, which was only linked directly to the Eastern Main Line
in 1911, was separated from Lourenço Marques by 620 kilometres of track
but was also a good 1700 metres (close to 6000 feet) above sea level.
Moreover, the difference in elevation was not evenly spread and could
therefore not be easily smoothed out over the longer distance. Indeed,
not far from the point where the bad-tempered little Elands River had
bitten its way through the rocky Escarpment to join the more composed
Crocodile River for the onward journey to Komatipoort, there was a particularly hellish section where there was a drop of around 200 metres
over just seven kilometres. It was the exceedingly rocky mountain terrain
that had to be negotiated between Waterval Boven and Waterval Onder
that may partly have informed the modern, modified lyrics of 'Shosholoza'
because it was at *that* hazardous juncture that returning Mozambican
workers experienced directly *kuleza ntaba* – the train running through

*The completed stone rail bridge over Dwaalheuwelspruit,
near Waterval Onder, 1893.*

the mountains – on their poignant journey to homes spread out over the largely barren coastal plain.

So severe was the drop between Waterval Boven in the west and Waterval Onder in the east, a distance of seven kilometres, that, for a decade and half, it could only be negotiated by imported locomotives of a type used in the Alps – ones that clung to a rack rail by means of a cog wheel or pinion – over a bridge or two and through a marvellous tunnel that still stands. Fraught and slow in equal measure, that section was eventually circumvented, in 1908, by running new tracks along the northern banks of the Elands. But, even then, bridges perched more than 20 metres above the river ensured that this part of the journey, more especially during the swifter west–east passage, remained potentially lethal.[9]

When viewed from a passenger's perspective, the long inward- and outward-bound hauls between Lourenço Marques and Johannesburg consisted of several legs, with each stage posing distinctive physical challenges irrespective of the quality of the rolling stock provided for transporting black workers. The run that seemed shortest, across the familiar hot, humid and swampy coastal lowland between Lourenço Marques and Ressano Garcia/Komatipoort was invariably uncomfortable. This commenced, or ended, with delays occasioned by the formalities and paperwork that documented a tetchy marriage entered into for money between the Portuguese and the South Africans. By comparison, the movement between the border and Waterval Onder was simple, but it, too, ended in a wait as the custom-built locomotives required to negotiate the sharp incline or decline that marked the passage through the Devil's Throat to (or from) Waterval Boven were hooked up to the coaches.

The relatively smooth onward ride from Waterval Boven to Witbank only flattered to deceive. The Highveld, seemingly innocuous when contemplated from afar, is perfectly Janus-faced, capable of being either mercilessly hot and dry or freezing cold and wet depending on a moody temperament dictated by sharply divided seasons that refuse to tolerate any lingering autumn or spring. Here, deep in coal country proper, there was another organised delay as those whom the fickle managerial gods consigned to collieries were separated from the large majority of workers who were destined for the gold mines some way down the line.

By the time the train left Witbank for the Rand, it was only too evident that the hazards that nature had tried to impose on the railway paled beside those left by the quest for gold. Between Germiston and Johannesburg it was man rather than God who had left his mark. The earth's insides had been ripped out, leaving behind mountains of rocky entrails and yellowing waste to bleach in the harsh sunlight. There was a dramatic increase in the number of tracks heading off in just about every

conceivable direction, and trains, especially those carrying freight, slowed
to almost walking pace as they gave way to those conveying more privileged urban commuters and workers.

For incoming black miners, regardless of whether the surroundings
had been sculpted by God or gouged out by man, not one of five stages
of the trek appeared to be in a hurry to give way to the next. For them the
long haul was a seemingly endless, buttocks-breaking, mind-numbing
confinement on steel rails stretching over two days. The closest thing to
an express service linking the Rand to Delagoa Bay in the period between
the formation of the South African Railways, at the time of Union (1910),
and the outbreak of the Second World War was the thrice-weekly mail
train. But even that supposedly upmarket service, catering almost exclusively for European passengers, took between 17 and 18 hours to cover
a bit more than 600 kilometres at an average speed of around 30 kilometres per hour at best – and that only when the system was allegedly
functioning at peak efficiency.[10]

During the interwar years, when successive South African governments
were increasingly committed to reserving better-remunerated positions
on the state-owned and -operated railways for the most poorly educated
and economically vulnerable of their supporters, white workers in the
system referred to the train carrying black migrants on the Johannesburg–
Ressano Garcia route in partly mocking terms as the 'Kaffir Mail'. It was,
of course, anything but a mail train – indeed, that was part of the joke.
The fact that a train conveying black miners was invariably slower than
regular passenger trains, and more often than not running late, only added
to the humour of those who did most to determine how long it would
take for the migrants to complete their journeys. In the early days, before
1910, even the train that terminated at Ressano Garcia – lopping 88 kilometres off the longer 620-kilometre journey to Lourenço Marques – took

26 hours. In 1936, the same run could still take 24 hours to complete even when the system was supposedly operating smoothly.[11]

But even those average travel times could be misleading and thrown out of kilter depending on the vagaries of the global economy or during wartime disruptions. To sense what did most to regulate the pulse of the system and the volumes of traffic generated, there was, however, little need to turn to the hidden hand of the market. It was there, in full view for all to see at either end of the system – the Witwatersrand mining industry and the port city of Lourenço Marques. And, of the two, it was clear which exercised a determining influence on the economy – the gold mines – and, consequently, on the train dragging migrants through the Devil's Throat around Waterval Boven.

# PART II

# CHAPTER THREE

# Mozambican Labour Regimes and the Eastern Main Line

## From Forced Labour to 'Market Forces'

*Our life is a constant journey, from birth to death. The landscape changes, the people change, our needs change, but the train keeps moving. Life is the train, not the station.*

— PAULO COELHO

With the exception of two short periods from 1919 to 1925, and then again from 1932 to 1934, when Britain temporarily abandoned the gold standard, the flows and strategic reserves of gold dominated global monetary systems and helped underpin the soundness (or otherwise) of the patterns of international trade until the 1970s. It was the decision of first Britain, in 1821, and then later the Federal Reserve System of the United States, in 1934, to purchase gold at a fixed price of $35 an ounce that did most to the underwrite the cohesion and stability of the international monetary system for most of the first half of the 20th century. These realities, determined by the economic power of capital spread around the countries of the North Atlantic rim, were, after the discovery of gold on the Witwatersrand, given further credence by an increasing supply of gold bullion drawn from South Africa, where a notoriously capital-intensive mining industry was always hopelessly dependent on a huge supply of ultra-exploitable cheap black labour drawn from within the country and across the wider region.

The extent to which the emerging global economic system was being fed by developments in the southern hemisphere was strikingly evident even before 1914. In 1890, the par value of capital invested in the low-grade Witwatersrand mines stood at £22 million, by 1899 it was up to £75 million, and by 1914 it was £125 million. In 1886, the infant Rand industry produced less than 1 per cent of the world's gold, by 1898 the figure had sprinted to 27 per cent, and by 1914 a whopping 40 per cent of the world's supply of the precious metal derived from this astounding, still developing new source.[1]

Much of this extraordinary transformation was accomplished with bursts of the creative, enabling and restless profit-seeking ability that capitalism is, and was, rightly famous for. Finance raised, controlled and distributed globally through a newly developed group system of mining houses, and the engineering marvels that came with deep-level mining realised on a scale never before contemplated, along with a new way of recovering and then refining gold from reefs of stubborn, unyielding quartzite, all emanated from the Witwatersrand.

Between 1887 and 1965, the new Eldorado in the far south yielded millions of ounces of gold valued at many millions more in pounds sterling, and provided shareholders in the United Kingdom and elsewhere across the developed northern world with a source of income that was almost without parallel not only by way of the size of the dividends it offered, but also in terms of its security. It was only after 1935 that investment in the Rand mines began to wane, but even then the returns remained impressive. Over the long haul, between 1887 and 1965, the industry paid out dividends that averaged 5.2 per cent, a rate greater than that provided by gilt-edged securities. From 1919 to 1963, the average yield was 9 per cent, as compared to 7.6 per cent for UK equities.[2] If there were investment 'risks', they were reasonably manageable.

However, creating wealth from consistent but extremely low-grade ore buried in reefs dipping away at increasing depths meant that, almost from the outset, the Witwatersrand mines demanded enormous quantities of capital and a near-limitless supply of cheap labour if they were to generate profits capable of retaining northern investors. In the civilised and developed world north of the equator, one supposedly filled with 'risk' for nervous Americans and Europeans investing large sums of capital in distant emerging economies, it was hugely reassuring if colonial governments and those managing the mines could, between them, attempt to control at least one element of uncertainty by doing their best to ensure that wages for unskilled black workers did not rise to the point where they eroded company returns. In the event, and throughout most of the 20th century, the southern African states, and more especially most of the mine owners – whose primary, if not sole, loyalties in terms of culture, lifestyle, wealth creation and choice of domicile upon early retirement lay in the northern hemisphere – did better than hold black wages; they managed, over 50 years, to lower them continuously.

Throughout the pronounced boom and markedly shorter bust cycles that characterised the industry in its foundational decade, that is, before the outbreak of the Anglo-Boer War, mine owners and managers attempted, with growing measures of success, to resist growth in black wages. After the war, in an industry increasingly centralised and better organised, one that used imported Chinese indentured labour (1903–1910) to sit out a short post-war increase in African bargaining power in the labour market for unskilled workers, these efforts were pursued with ever greater focus and purpose.[3] Between 1896 and 1911 the employers' organisation, the

Chamber of Mines, reduced black wages as a percentage of total costs in the industry by more than half, from 25 per cent to 11 per cent. The years immediately after the First World War saw a 50 per cent increase in the cost of living on the Rand but no commensurate adjustment in wages for hard-pressed black workers. Nor were the two decades that followed any more promising for supposedly unskilled miners, even though in practice there was evidence of their growing competence and increasing levels of productivity. The squeeze was remorseless. The cash wage bill for African miners fell from 13.7 per cent of total revenue in 1936 to 8.9 per cent in 1969: 'In real terms, using 1938 as the base year, black cash earnings in 1969 were no higher and possibly even lower than they had been in 1911.'[4]

The sustained assault on the declining earnings of African mineworkers drawn from southern Mozambique did not end there, however; it was taken further. In 1940, amid the financial deprivations of the Second World War, the governments of South Africa and Portugal agreed that the deferred payment of wages due to black miners upon their return to Mozambique might be paid in gold bullion at a fixed price that the two countries had agreed upon. Given Johannesburg's expertise and knowledge of how the international bullion market worked, the Portuguese government allowed the South Africans to sell the gold on the world market and then transfer the proceeds to the colonial administration in Mozambique.

While the wages of black Mozambican labour continued to decline, or at best remained relatively stable in real terms, the difference in the price of gold between the official rate and what it fetched on the free market continued to widen. This meant that by the early 1970s, and right up to 1977, when the scheme eventually ended, 'Portugal's gold holdings made a high – often spectacularly high – profit from the gold sales.'[5] None of the profits realised from this protracted and cynical arrangement accrued directly to the African miners of the Sul do Save. In effect, a state that had coerced and sold colonial captives into industrial servitude at ultra-cheap rates in exchange for guaranteed rail traffic to its principal port city then went on – with the active connivance and support of a neighbouring state – to intercept and further reduce the wages of those working in the Witwatersrand mines.

Here, then, was an industry successfully defying the laws of economic gravity. Despite real wages declining for more than a half century, the Rand gold mines – except for a few periods of usually short duration – not only grew, but also managed to maintain their complement of black workers. It was an extraordinary achievement that reflected not only near-untrammelled industry and state power at the point of production itself, but also the increasing hardship and poverty in those rural areas pushing out migrant labour, as well as an ability to transport and deliver those caught in the double bind to a grossly skewed, largely captive, labour market on the Witwatersrand. By the mid-1950s the original 'pull' of black wages for Africans had long since given way to 'push' factors in the southern African countryside, and beyond that, in rural southern Mozambique, to forces that were more insidious still.

The number of black workers required by the gold mines over half a century varied from around 77 000 in 1903 to some 200 000 by the late 1920s. It then rose consistently to over 300 000 in the 1930s and stayed at those levels until at least the mid-1950s and beyond. Only once, in 1908, did that number ever fall below 150 000. And within those formidable numbers, not once over a 12-month period was there a time when the majority of miners were *not* drawn from Mozambique, and within that from the Sul do Save.[6] As a Portuguese politician put it in 1922, the southern part of the country had, in essence, been reduced to 'one large compound for natives of the Rand'.[7] In the same year, in South Africa, the dedicated recruiting arm of the Chamber of Mines could trumpet the success that it had enjoyed in institutionalising a low-wage regime in the Sul do Save:

> The mining industry has been built up largely upon the labour of the Portuguese natives, who worked underground when no other natives would do so. Mining work is now second nature to the Portuguese native. He has become a 'natural miner', and he goes to no other industries, and his whole life is centred round the custom of work on the Witwatersrand gold mines.[8]

The dividend, paid in ultra-cheap black labour, was spectacular. From the founding of the Witwatersrand mines until 1910, southern Mozambicans made up more than 50 per cent, sometimes 60 per cent, of the African workforce, a figure that only began to fall in the 1930s. As late as 1955, men drawn from the Sul do Save still comprised more than 30 per cent of the gold mines' black labour complement.[9]

For most of the 20th century the Chamber of Mines and the highly focused cartel that steered the fortunes of its recruiting arm, the WNLA, crowed publicly as well as privately about how its primary source of cheap labour was, for the most part, Africans from Mozambique, who had happily entered into contracts significantly longer than those customarily agreed to by black South Africans through a sister organisation, the Native Recruiting Agency (NRC, established 1912). There was, said the chairman of the WNLA, referring pointedly to that part of Mozambique that lay below 22 degrees south latitude, at his association's annual general meeting in 1913, no gainsaying 'the popularity of mine work among these natives'.[10] It was, if chairman John Munro were to be believed, a serendipitous arrangement, one in which news of the declining value of wages, calculated to the last decimal point by the Chamber of Mines, was readily embraced by tens of thousands of Africans who, collectively, showed little or no understanding of how regional transnational labour markets functioned. In fact, the opposite was true.[11]

The facts were easy to determine. There had been no mass brain failure among black migrants drawn from south of the Save. Colonial states throughout the region – including in Portuguese East Africa, where there were no great extractive mineral industries of note – had rigged domestic labour supplies in favour of the mine owners. In practice, the overwhelming majority of southern Mozambican males – Thonga, or so-called Shangaans – had virtually no choice other than to proceed to the Transvaal on foot or by rail. It was not so much the 'pull' of a free and fair labour market characterised by declining real wages that 'attracted' African migrants to the Rand mines, but rather the appalling alternative of having to undertake forced labour – *shibalo* – for even lower wages, or sometimes none at all, on state or private projects favoured by a corrupt and violence-prone Portuguese administration. Given the choice between labouring under near-slave conditions on your doorstep for no, or ultra-low, wages and industrial servitude at higher cash wages that were declining in purchasing power in a neighbouring country, toiling in the gold mines indeed became a 'popular' choice – or, shall we say, one of the few sensible choices available.[12]

*Shibalo* was the bastard child of two longer-lived and only recently deceased systems – slavery and indentured labour – that had ravaged the bodies and minds of African men and women over several centuries. Portugal was among the earliest European nations to engage in the global slave trade. Despite entering into agreements to end the trade, in 1815 and 1836, it only formally ended human trafficking in 1878, making it one of the last nations to do so. 'Indentured labour', frequently little more than slavery masquerading as freely-entered-into contract employment recorded on shipping manifests, was, in effect, merely a bridging device to facilitate a slow, quasi-legal exit from the slave trade.[13] In Mozambique, most capturing of slaves and 'recruiting' of indentured labourers took place in the far northern half of the colony, but these activities left long-lasting scars on the consciousness of Africans up and down the coast. As the American consul, Stanley Hollis, noted as late as 1894, when a Portuguese navy transport put into harbour in the capital (until 1898), on Mozambique Island, 'the cry of "slave ship" goes round among the natives and there is a regular exodus of natives from the island to the continent'.[14]

Although undoubtedly not exposed to the worst excesses of the trade in slaves and indentured labour, as experienced in the north, the Sul do Save was not exempted from such trafficking. There was a fully fledged slave station at Lourenço Marques in 1820, and active slaving continued as far south as Inhambane and Delagoa Bay well into the 1850s and, in much diminished form, for possibly two decades thereafter, until its formal abolition in 1878.[15] The importance of this lies in the fact that legalised slavery in Mozambique overlapped, however briefly and imperfectly, with the emergence of the first large-scale mining operations on the Kimberley diamond fields and the earliest gold mines of the eastern Transvaal.[16] The experience of earlier and harsher Portuguese labour regimes was not so distant as to be entirely lost on either those doing the recruiting for the Witwatersrand mines during the first two decades of the 20th century, nor on those being recruited. Indeed, in some cases they were scarcely beyond living memory.

So closely did the previous slave and indentured-labour regimes resemble the recruitment practices of European agents for the massive new Rand gold-mining industry that their agents slipped easily into terminology characteristic of the older orders; they spoke of 'blackbirding', 'hunting' and 'purchasing' – words 'embedded in the lexicon of slavery'.[17] Likewise,

44

for Africans throughout the region the words *chibaro* or *shibalo* lay on a continuum of exploitation in which contract labour, forced labour and slavery were indistinguishable.[18] In short, the ghostly imprimatur of slavery, so recently ended, and the advent of its close relative, *shibalo*, helped underscore the 'popularity' of the Rand mines. If, in the far north, the arrival of a steamship could still set off the cry of 'slave ship', then in the deep south it was the steam locomotive that 'cut to pieces' and 'smashed' the men of Gaza, who as recently as 1895 were still fairly independent 'warriors', and threatened to whisk them away to nearby servitude.

There were good reasons why *shibalo* cast such a long shadow over the largely infertile Sul do Save. One of the primary objectives of all colonial administrations was to mobilise indigenous peoples as a source of cheap wage labour that could be utilised for developing the infrastructure of a weak state or private enterprise, and to raise badly needed revenue from a black working class it was intent on creating by way of taxation. Throughout most of the first half of the 20th century, however, the Mozambican administration was bureaucratically inept, militarily weak and more often than not on the verge of bankruptcy. With little support forthcoming from a numerically insignificant class of impoverished settlers, who were as capital-starved and skills-deprived as they were resentful at having to compete with corporate or state institutions for access to cheap African labour, the tasks of mobilisation and taxation were rendered even more difficult than they were in other colonial settings. In southern Mozambique, then, with the notable exception of the port and rail complex at Lourenço Marques, so weakly developed was the carrot of wage labour that the administration had to rely on the stick of forced labour to try and stimulate development in the face of far stronger 'market forces' radiating from South Africa. In parts of southern Africa there was, during the great age of primary industrialisation, often far more 'force' than

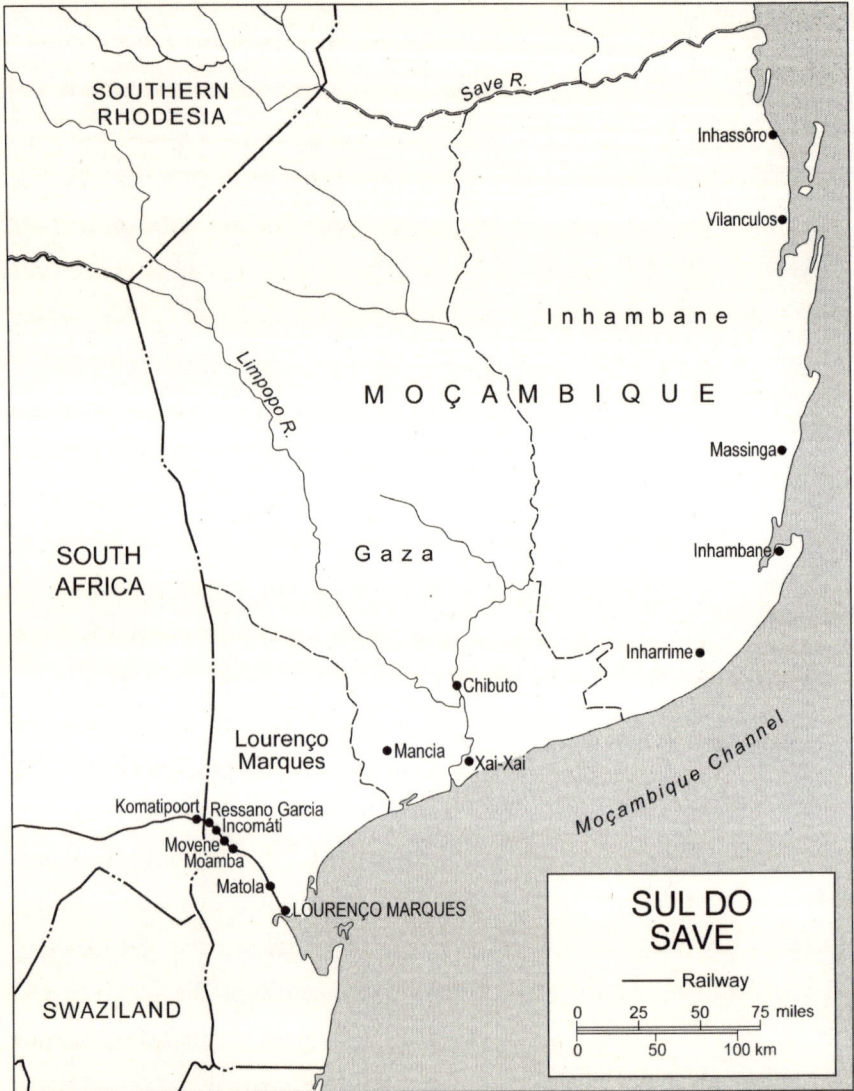

'market' behind the term 'market forces' in what is best understood as a single regional political economy with a strong centre dominating weaker peripheral nodes.

The system of forced labour for indigenous peoples, as implemented in the Sul do Save, may have had its origins in other parts of Portugal's empire but, when seen from a Mozambican perspective (and therefore possibly slightly misleadingly), it coincided nearly perfectly with the demise of slavery and the rise of southern Africa's primary extractive industries. The first sign of things to come from the Portuguese was a regulation, in 1881, making provision for compulsory labour to curtail any 'vagabondage' arising from the abolition of slavery.[19] But, more revealing may be the timing of legislation that came after the surge in profitability marking the end of the first decade or so of gold mining on the Rand.

In 1899, the Portuguese government, nominally Catholic, approved umbrella legislation that included Mozambique – ever more reliant on the economic fortunes of its powerful Protestant neighbour – that might as well have been drawn directly from the Calvinist teachings that had long dominated South Africa.[20] Self-sufficient indigenous people, hitherto wholly reliant on subsistence farming on often marginally productive agricultural land, henceforth had a 'moral and legal' obligation to find waged employment. If they failed to do so, they could be compelled to do forced labour on such projects as might be determined by the local administration.[21] God, it seems, was seldom wholly at odds with the emerging cheap black labour markets throughout greater southern Africa.

The 1899 Portuguese regulations, reminiscent of the 18th-century regulations governing corvée labour in French colonies, were amended in 1911 and 1914. They came at a convenient moment for the Rand mining industry, when it was hard pushed to extend its supply of black labour at steadily declining wages. These changes, however, left the law 'hard at the core' and only 'a little softer around the edges'.[22] Ironically, the use of corvée labour in Portugal's colonies did not decline when a republican administration replaced the monarchy in Lisbon in 1910 but instead actually increased.[23] In 1928, the code was again adjusted, removing the legal, but retaining the 'moral', obligation of indigenous subjects to seek waged employment or face the consequences of forced labour, which frequently meant working for months on end for no wages at all.[24] Yet again, this time fortuitously, the revision came on the eve of a period marked by the expansion of the Rand mining industry – a few years before South Africa abandoned the gold standard. As with the painfully slow exit from slavery, forced labour in Mozambique attracted growing international criticism as the century wore on, but, as before, the metropolitan political elite in Portugal pointedly refused to ratify the International Labour Organization's Forced Labour Convention of 1930 until 1956. In new disguises, coerced labour lingered on in the colony – including in the Sul do Save – where 'market forces' had almost entirely destroyed the always-precarious subsistence agriculture and ushered in a near-universal dependency on cash wages.[25]

As elsewhere in the colonial world, forced labour performed a bridging function within the Mozambican economy. As one apologist for the system, a white South African socialist, put it in 1933, 'from the commencement it was obvious that the native had to be slowly educated to the enormous benefits obtained by work in the mines'.[26] *Shibalo* helped span the years between the demise of a labour regime dependent on slavery and the emergence of a more modern economy, one in which black subjects from traditional forms of non-market, subsistence agriculture had, through taxation, became available as a large-scale source of cheap wage labour. During this painful shift, forced labour came to the rescue of a weak administration and economy in other important ways: it guaranteed unwaged labour for basic infrastructural development and maintenance; it contributed to Portugal's 1916–1918 war effort; it provided ultra-cheap or wage-free labour during the Great Depression; it facilitated forced peasant production of cotton and rice for a mercantilist market in metropolitan

Portugal between the two world wars; and it helped keep cash wages artificially low in the only meaningful local labour market – Lourenço Marques, with its vital port and rail installations.[27] At least as important, however, was the fact that forced labour helped underwrite the allegedly completely 'free' and 'popular' flow of black labour to the Witwatersrand mines and the even more exploitative collieries of the Eastern Transvaal, centred on Witbank.

Historians, regardless of ideological persuasion, generally agree that either the enforcement or the ever-present threat of *shibalo* labour within southern Mozambique during the earliest decades of the 20th century primed the pump for the WNLA's recruitment of black workers for the South African mining industry.[28] What is perhaps less well appreciated, however, is just how so tightly iterative a process became part of a much longer-lasting, self-reinforcing, chicken-and-egg system and therefore contributed centrally to the maintenance of ultra-low wage regimes on both sides of the border for more than half a century. The more the Portuguese administration resorted to the use of forced labour for public works or to raise taxes, the more 'attractive' the better, albeit declining, cash wages offered by the South African mines became, and the greater the resulting exodus of black migrants to the Witwatersrand from the Sul do Save. This increased the need of the Mozambican authorities to mobilise yet more *shibalo* labourers, thereby only increasing the outflow of workers to Johannesburg in a never-ending cycle of impoverishing, wage-lowering initiatives that benefited only the employers on both sides of the border.

Any criticism of so ethically reprehensible an economic dispensation was always vigorously denied or deflected by the authorities on both sides of the border. For the Portuguese, struggling to mobilise capital and labour alike, the problem lay in the loss of 'their' indigenous workforce to a bullying and wealthy neighbour. For the South Africans, accusations about the systematic abuse of Africans during 'recruiting' operations in Mozambique were simply a by-product of the well-known moral deficiencies of the Catholic Portuguese, only recently weaned off their liking for slavery. Yet, despite muttered misgivings about the informal functioning of the new system, Britain and Portugal had long since entered into bilateral agreements in which the labour of most African men, including those from south of the Save, had been traded in exchange for a percentage of the rail traffic carried along the pivotal Eastern Main Line.

In 1897, the Chamber of Mines, enjoying access to the recently completed Delagoa Bay line but as yet unable to rely on the backing of a state under the control of the Kruger government, made a largely unsuccessful attempt to monopolise the recruitment of black labour in Mozambique. A British victory in the Anglo-Boer War, followed by successive compliant administrations in a hurry to entrench the power of the mine owners before the advent of Union in 1910, strengthened the hand of the Chamber of Mines immeasurably. Lord Alfred Milner, overseeing post-war reconstruction in the newly acquired Transvaal Colony, gave his support almost immediately, via a modus vivendi entered into with the Portuguese in 1901, allowing the WNLA exclusive, privileged access to black labour across Mozambique in return for a guaranteed volume of rail traffic routed through the port of Lourenço Marques. It was, as noted by one historian, 'a naked exchange of rail service for human service' in an agreement never consented to by African workers.[29]

Thus, at about the same time that Edmund Morel and Roger Casement were breathing life into the Congo Reform Association so as to challenge the savage, arm-lopping, rubber-collecting proclivities of King Leopold's Congo Free State, the British and Portuguese political elites agreed upon a transnational arrangement for the better-concealed exploitation of Africans in the coal and gold mines of southern Africa via a modern form of industrial servitude that consumed black lives almost as readily as it did limbs.

The Portuguese sold the African men and boys of the Sul do Save and their birthright to seek fair and equitable treatment in the regional labour market to the British imperialists and their Witwatersrand mining-house allies. And, just as surely as the biblical Joseph was sold to human slavers and traffickers by his brothers, so in that same moment was born a train of the Midianites that would ferry the males of southern Mozambique to and from South Africa.

With the modus vivendi serving as template, the South Africans and their Portuguese counterparts used the post-war reconstruction era and what followed to gauge the relative strengths and weaknesses of their partners. The immediate objective was to leverage any developing advantage in ways that might allow for a sharpening of economic focus so as to extract further financial concessions at the expense of those who would have no voice in the subsequent negotiations – Mozambican Africans being forcibly pushed off the land in order to create a class of migrating workers and avoid incurring the costs of maintaining a permanently resident, decently housed, properly fed or educated resident black proletariat on the Witwatersrand. The Eastern Main Line played a small pioneering role in undergirding the region-wide industrial revolution that rested first on segregation, and then, much later, on apartheid.

The Mozambique Convention of 1909 guaranteed that 50–55 per cent of the rail traffic destined for the 'competitive zone' circumscribing the Witwatersrand would pass through Lourenço Marques in return for the Sul do Save – well within the penumbra of state-sanctioned *shibalo* raids – being turned into an exclusive labour recruiting zone for the WNLA.[30] That agreement was followed, almost immediately, by the passing of Halley's Comet, which, while it might have lit up prospects for enhanced profits among the few, may have been seen only as an ominous development by the many experiencing the forced 'march of progress'. The 1909 convention, despite occasional appeals by the parties for minor adjustments, served the signatories reasonably well for the better part of a decade. The First World War, however, marked the end of an era in the Sul do Save in which *shibalo* labour, acting as the priming mechanism par excellence for successful WNLA 'recruiting', very slowly began to make way for a far greater flow of 'voluntary' labour for the mines. This was unleashed by 'market forces' driven by the progressive decline of a rural economy systematically deprived of its male labour, the progressive enforcement of a punishing domestic tax regime, and the resulting, growing 'preference' for cash wages.[31]

By the early 1920s, circumstances on either side of the border had changed sufficiently for the signatories to the original convention to want to renegotiate the agreement in ways that took cognisance of the emerging post-war realities. South Africa, having put on economic and political weight after the advent of the Union, and having had its territorial ambitions for regional expansion whetted by wartime military successes, aspired to outright control of the port of Lourenço Marques if not its wholesale incorporation.[32] Portugal, by contrast, had been crippled by wartime debt and was mired in a series of currency and financial crises that helped pave the way for a military coup in 1926. This eventually ushered in António Salazar's *Estado Novo* (New State), whose agricultural and mercantilist policies aimed at making the colonies pay their own way – and, whenever possible, that of the metropole as well. For the latter to happen, however, would require significantly more effort to develop the agricultural and commercial prospects of Mozambique and other colonies with minimal capital and an insignificant settler presence. Developments in the slowly emerging agricultural – as opposed to plantation – sector only prolonged the life of the detested *shibalo* labour system.

The successful renegotiation of the convention in 1928 – it was renewed in 1934 and 1939, and lingered on in phantom form into the 1960s – reflected a growing confidence on the part of the Transvaal Chamber of Mines and the appearance of the first, although still extremely weak, economic muscle of the Mozambican administration.[33] The South Africans, keen to ensure that recruiting practices in the Sul de Save were publicly seen to be ethically based, were more trusting of the pull of 'market forces' to deliver them the supply of black miners that had previously relied heavily on the push of *shibalo* labour. The initial construction of a cheap labour market in a cross-border colonial setting had necessitated a good deal of dirty work at the crossroads, but once the capitalist template was securely in place, clean-up operations could commence – provided that black workers continued to be denied the legal right to organise recognised trade unions at the points of production in South Africa.

The 1928 convention saw an end to the WNLA's recruiting monopoly in the southern part of the country and capped the number of migrants at 100 000 per annum – a number that was thereafter to be reduced by 5 000 a year for four years. Initial contracts for mine labour were set at a maximum of 12 months, although provision was made for a further extension of six months. Any Sul do Save recruit who failed to return home after the expiry of his contract could be legally excluded from the Union. These concessions were meant to appeal to disgruntled Portuguese settlers in search of their share of cheap labour, while local merchants were, after a decade-long struggle, to benefit from a deferred pay system that would ensure that part of migrant earnings would be spent back in Mozambique. Finally, while an Advisory Board would oversee the port and rail complex at Lourenço Marques, no direct administrative intervention into the management of the system there would be allowed, and the volume of traffic to the competitive zone on the Witwatersrand – utilising the Eastern Main Line – would remain unchanged.[34]

This gradual diminution in the influence of the Chamber of Mines, the reach of its recruiting arm and the collective squeeze on black labour and wages came only after three decades during which *shibalo* brutality and WNLA exploitation had marched in lockstep, with the latter determining both the direction and pace of operations in a war for workers. The WNLA was for many years, as one Portuguese administrator noted, in effect virtually a 'state within a state' running a 'parallel administration' that accounted only to itself.[35] In 1906, the association employed over 1 200 agents across southern Mozambique, with its 'runners' dressed in the same distinctive red uniforms used by the *sipais* – the 'native police' responsible for turning out *shibalo* labour.[36] In the Sul do Save, the WNLA had more personnel on the ground than did the colonial administration. In 1911, in a formally commissioned inquiry partly of its own making, WNLA officials claimed that 'at least 75 per cent of the boys secured through the organisation offer themselves voluntarily at the camps and stations of the WNLA'.[37] How the other estimated 25 per cent – one in four 'recruits' – found their way to the association's recruitment stations was perhaps best left to the imagination, perhaps not. The entire operations of the mining industry, suggests a leading analyst, were 'quasi-military' in their organisation.[38] From 1910 onwards, the South African English-medium press carried consistent reports and rumours of slave-like recruitment practices in Mozambique by any or all organisations bar, of course, the WNLA itself.[39] Many WNLA recruits from the Sul do Save referred to themselves as having been 'sold' to 'Mzilikazi', the name given to both the WNLA itself and its reception compound in Johannesburg by most men living in southern Mozambique. It was not the most flattering of monikers, since the name of the great Ndebele warrior-chief who had wreaked havoc across the southern African Highveld in the mid-1830s was synonymous with fear induced by bloody invasion, forcible dispersal of the vulnerable and a good deal of slave-taking.[40]

From all of the above, it would seem that the mobilisation of black labour for the South African mining industry during the first 50 years of the 20th century was the by-product of direct, albeit diminishing, physical coercion at source in the Sul do Save, and that, in the decades that followed, the mines continued to rely heavily on voluntary labour derived from market forces primed by progressive rural underdevelopment in southern Mozambique – something that, at least in part, was itself a consequence of the earlier enforced migratory labour. The circulatory of this experience within an oppressive regional regime was, however, not confined to the increasingly poverty-stricken homesteads of African peasant-proletarians in the Sul do Save nor, as we shall see shortly, to the mine compounds of the Witwatersrand. The return cross-border rail journeys undertaken between Mozambique and South Africa were just as reflective of a coercive, exploitative and self-reinforcing transnational system as any experiences undergone at its terminal points. It was the formidable middle rail passage that linked the two.[41]

Before recounting more fully the lived experiences of African men on the rail journeys between the Sul do Save and the Witwatersrand, it may be helpful to pause and reflect briefly on how those pivotal black movements were shaped by the colonial ideologies and managerial perceptions of the two parties most immediately responsible for running the rail system: the junior European lieutenants of the rapidly developing gold-mining industry and the everyday white operatives of the South African

Railways. How, then, were men drawn from a region still oozing a painful history of forced labour and slavery, who enjoyed virtually no economic, social or political rights, to be classified, transported and initiated into a capitalist system as workers-in-the-making?

# CHAPTER FOUR

# Colonial Managerial Ideologies

## Categorising Black Men in Transit

*When the train of history hits a curve, the intellectuals fall off.*
— KARL MARX

The idea behind the Eastern Main Line, as already noted, predated the discovery of the Witwatersrand goldfields, and the railway was meant to provide the small Boer agrarian elite who controlled the Highveld with a rail link to a port that was beyond direct British political control. The ambitious foundational thinking of the 1870s was, however, so far removed from economic necessity and realities that it would slumber on through promising developments in the Eastern Transvaal. The discovery of gold around Johannesburg, in 1886, changed everything. Instead of the rail link being called upon to serve farmers in search of new markets, it would now serve a mining industry located in a fast-developing urban economy. This provided the projected rail system with a new determinative logic: to provide an efficient way of importing heavy machinery, food and other merchandise at competitive rates and a minor export route for such few agricultural and other products as a farming economy fast giving way to an industrial order was able to muster. Regardless of who 'owned' or oper-ated the Eastern Main Line after 1886 – first the NZASM, then, after the war, the British-controlled Central South African Railways (CSAR), and thereafter the SAR – the rationale, the preferences and the priorities of the management were always to favour goods over passengers.

Right from the outset, however, the Chamber of Mines, and more especially so after it launched the WNLA in 1902, had a dual dependency on the emerging rail system. It required not only machinery, timber and other supplies but also an ever-growing number of men to be moved along the Eastern Main Line – the thousands of African migrants drawn from its preferred pool of ultra-cheap East Coast labour agitated by the stick of Portuguese forced labour. The dedicated WNLA trains bringing workers in from Mozambique and returning them once their contracts had expired therefore occupied a somewhat anomalous position as passenger trains in a rail system designed and operated in such a manner as to give preference to freight. The day-to-day operations of the Eastern Main Line were thus beset by an underlying conflict arising largely from fundamentally differing managerial priorities, with the SAR wishing to give preference to the movement of freight and the WNLA officials wanting to prioritise the movement of men in and out of the labour-devouring mining industry.

Problems of this nature, calling for accommodation and compromise in determining logistical preferences when moving freight and passengers through a system, were not peculiar to the Eastern Main Line. Indeed, they were – and probably still are – a feature of railway management in most industrialised societies. In colonial settings dominated by racial thinking, however, what was a universal problem assumed an additional, complicating, local dimension that exacerbated tensions between the managing authorities of the system.

In the case of the SAR and WNLA, the men being transported *en masse* were no ordinary men – citizens drawn from different classes with distinctive political voices seeking incorporation into an emerging shared socio-economic order. Far from it; 'they' were migrants, and foreign black migrants at that. 'They' were recently conquered Thonga-Shangaan warriors with real and imagined resentments, untutored subsistence farmers being moulded into a labour force for a new industry in an emerging capitalist order that tended to treat workers first and foremost as a collective, as a commodity, as a factor of production and only thereafter as individuals with distinctive characters, temperaments and histories.

In short, the unwritten underlying codes of management for both the mining industry and the railways cast migrant workers primarily as commodities – as things – and only secondarily as men undergoing their first experience of discipline, order and time, not as it flowed from the seasonally predicated economy they were accustomed to, but as part of a modern industrial order in which the clock and the calendar governed the rhythm of a year. But, while railway managers were almost always comfortable with viewing trains carrying migrant labourers as consisting primarily of inanimate freight, of things that required costs to be recovered in appropriate ways rather than as part of an authentic passenger service, in practice this gave rise to problems of comfort, health and safety that could become sufficiently counterproductive for WNLA officials to raise questions of compassion and humanity when dealing with their undeniably human charges. The railway officials treated men as freight, but the WNLA – never shy of treating labour primarily as a dehumanised commodity when convenient – could not afford to allow a too-slow freight service to degenerate to the point where the health or safety of its charges compromised the economic logic or smooth functioning of the mining industry as a whole. Thus, at various moments, and with varying degrees of enthusiasm, the WNLA pushed to have the trains treated as passenger trains, enjoying the attendant privileges of priority when passing through the system.

Reports from the SAR and WNLA covering trains ferrying African 'recruits' and 'repatriates' between Mozambique and South Africa over the first half of the 20th century are noteworthy for the pervasive use of terms that served to conceal the true status of passengers destined for insertion into an industrial regime that was becoming ever more oppressive and racially segregated for people of colour and by class. Presumably so as to distinguish them from the single most powerful block of politically active industrial workers in the country between the two world wars – the militant white miners who, in 1922, brought the country to a standstill in an unsuccessful 'revolution' – black recruits were called anything but 'miners'. Any such recognition might have undesirable implications when it came to the acquisition of skills or proper payment for vocational callings, or debate about permanent residence. The best that 'recruits' could hope for was to be called 'mine natives', a formulation that gave priority to the incumbent's temporary place of employment in the system rather than to the individual and his occupation. And, in keeping with thinking that served to infantilise men who until recently had been independent subsistence farmers or warriors, Mozambicans, or 'foreign natives' – to use an oxymoron much favoured by white South Africans – were most often cast as 'East Coast Boys', a small step up from the still more pervasive 'boys'.

This awkwardness – in finding the right words to define men-boys and foreign-natives – extended to all levels of the SAR when it came to classifying the WNLA trains moving miners to and from the Witwatersrand. If the trains were carrying mere itinerants, blacks of uncertain status and unspecified temporary occupation moving between far-off homesteads and various mines on the Witwatersrand or in the Eastern Transvaal, they did not qualify unambiguously as passenger trains carrying people readily identifiable as workers. Yet, for all that, the SAR was forced to acknowledge that the trains *were* transporting an essential factor of production, massed cheap labour, to and from the most profitable industry in the country – the gold mines. Whatever formulation the SAR senior administration therefore came up with reflected these ideological tensions.

The annual reports presented to both Houses of Parliament by the General Manager of the SAR itemised the volumes of passenger traffic within the system as carried by WNLA trains at specially negotiated bulk rates. These reports groaned beneath the changing times and the political weight of appropriate classification. In 1910, WNLA trains were reported on under the subheading 'Natives in Batches' – 'batch' being a term the *Oxford English Dictionary* defines as being a 'quantity or consignment of goods produced at one time'. Incidentally, for much the same reason, when labour was seen to be flowing freely from the Sul do Save, as it did in, say, 1927, the WNLA District Manager in Lourenço Marques wrote of recruiting operations as if they were part of factory-like production, pleading for a second train 'while the output remains high'.[1] From 1913 to 1947, the traffic was accounted for in SAR reports as 'Natives by Goods Trains'. It was a formulation that accurately portrayed the frequency with which freight-carrying bogies were hitched to what supposedly were passenger trains and therefore mutated into the downgraded category of 'mixed trains' when it came to scheduling priorities. For much of the decade that followed there was a painfully slow shift from focusing on the nature of the rolling stock to the occupants of the coaches themselves: the WNLA trains then came under the subheading 'Special Native Fares', which by 1958 had become 'Third Class Fares', and then, with the coming of the grand insanity of systematically applied apartheid in the 1960s, 'Special Bantu Fares'.[2]

Lower down in the SAR and WNLA managerial hierarchies and, more especially, at operative levels, the tensions in the terminology that arose when trying to find words capable of either recognising or imposing differences between goods and people, the inert and the living, men and machines, were extremely stark. The use of the word 'waybill', customarily referring to a list of either goods or passengers, was not in itself particularly revealing of how Mozambican miners on the move were seen or treated. But when the same migrants were also spoken of as having been 'imported', or as qualifying as 'human freight', it becomes clearer that the migrating miners were far closer to being perceived as chattels – items of property – rather than as people.[3]

The mining industry itself was a central force when it came to choosing language that would shape the way inexperienced rural black migrants from the Sul do Save would be classified and treated on their journeys into the new world of industrial labour. In the earliest years of the 20th century, when the military advance of colonialism and frontier wars still figured prominently in the minds of white managers, black Mozambicans were cast as 'nothing but raw savages' who, over time, became 'raw natives' before graduating, in the 1950s, to the comparatively refined status of 'primitive natives'.[4] This grudging recognition of the migrants' standing as men, albeit ones drawn from a strange and threatening universe, hovered on the edge of downgrading them from human to animal status.

Migrants, suggested one WNLA official, in a private note to the state's Director of Native Affairs in 1911, knew what to expect once they had been recruited: 'They go up to the Rand and return throughout each year and form a moving circle, and, like circus animals, require whipping up at points around the ring.'[5] A contractor tendering for WNLA business in 1917 offered a choice of five different types of biscuits for possible black consumption: mine biscuits, superior mine biscuits, a nut biscuit for mine use, a 'European white Biscuit' (containing 15 per cent meat and fat) and a 'Dog Biscuit'. In similar vein, catering for mine recruits was undertaken to 'enable feeding', while those migrants not readily classifiable or easily traceable were occasionally cast as 'stray natives'.[6]

On the Eastern Main Line this subliminal elision of animal and human categories was taken to its logical conclusion and manifested itself in startling physical form in the composition of WNLA trains when the British administration took over the management of the system after the Anglo-Boer War. From the end of the war and for more than a decade thereafter, tens of thousands of migrants were regularly transported in open cattle and/or coal trucks of the CSAR.[7] In 1910, the new SAR administration made a half-hearted attempt to modify and modernise the 160 bogies that had, most reluctantly and slowly, been provided with the crudest possible ablution, drinking-water and sanitary facilities over the preceding period.[8] The WNLA management, somewhat keener to prioritise the human over the animal dimension of the movement of contracted migrant labourers, was not persuaded by either the quantum or the quality of the improvements effected to bogies by the state:

> Of this number [160], 58 were converted Central Government Railway [CGR] cattle trucks, 13 native box trucks, and 89 were CSAR native coaches of the type agreed upon for our service. You will therefore see that 71 of the 160, or say one-half of the coaches supplied were of a type well recognised as unsuitable for our requirements.[9]

While the situation with regard to the quality of accommodation provided on the Ressano Garcia run improved slowly after the introduction of third-class 'native coaches', an insufficiency in the number of suitable passenger coaches remained a feature of the system for at least two more decades. Throughout the 1920s the Chamber of Mines and the WNLA continually complained to the General Manager of the SAR about

Mozambican repatriates having to travel from Johannesburg to Witbank, or from Witbank to Waterval Boven, in open coal trucks in autumn and late winter.[10] It is unsurprising, then, that the most focused of those historians tracing the capacity, as well as the willingness, of the national rail system to cater for vulnerable black men being moved to and from the Witwatersrand over great distances came to the conclusion that 'Migrant workers were often treated as animals, or worse, as pieces of cargo' – as indeed they were.[11]

Designating the train carrying recruits – half-animal, half-human – to the mines as a 'labour train', as the 'Kaffir Mail' or simply as a 'kaffir train' did nothing to lift the classification of migrant labourers from the category of freight to passenger.[12] While English-speaking WNLA officials, more urban-based and literate, were content to document their perception of the train in those terms, less well-educated SAR employees – many of them semi-literate Afrikaans-speakers recently off the land, and given preference as 'poor whites' between the wars – implicitly coded almost all trains conveying black workers as 'goods trains' or, when coal or other freight trucks were added to the train, as they frequently were, as 'mixed trains'. This meant that for almost the entire period, between 1910 and 1960, the comfort, safety and speed of the trains carrying migrants was determined largely by lowly white operatives who, as vote-seeking politicians constantly reminded them, were vulnerable to the 'unfair' competition for labour represented by these black workers.

Refusing to treat trains carrying Mozambicans to and from the mines as passenger, rather than as freight, trains could – as we shall have occasion to note in due course – have fatal consequences, not least of all because only more experienced, higher-grade locomotive drivers were meant to operate the passenger trains that negotiated the system at strictly regulated speeds. Less dangerous, but more irritating for WNLA officials, were the chronic delays in the arrival times on the Witwatersrand, which would leave migrants angry, exhausted and hungry. Delays bedevilled the mass bureaucratic processing and medical checks that recruits had to undergo at the WNLA depot in Johannesburg prior to being distributed to the mines. The persistent refusal of SAR officials to concede that incoming trains were conveying people rather than goods made for enduring frustrations at the WNLA. 'The natives get restless,' complained a white conductor in 1941, because of an unwillingness on the part of the SAR to facilitate the movement of WNLA trains through the system as passenger trains. 'The remedy appears to be,' he argued, 'to insist and keep insisting until the train is treated as a passenger train, between Witbank and Booysens, as apparently under present conditions we are at the mercy of signal cabin officials.'[13] The pressures associated with wartime traffic would not have helped, but it was a long-standing objection and one that never quite went away.[14] A decade and a half later, the same old refrain was still being heard. In 1955, an irked conductor noted that incoming trains – by then numbered 830 – were treated neither as 'mixed' nor as passenger trains but run instead 'more like a circus train as all goods trains are given preference to the 830 from Witbank to Booysens'.[15] One can only guess as to what part of the circus he might have had in mind when offering that pointed observation.

The insidious ideologies that SAR and WNLA managers used when framing the classification of migrant trains, or when describing the status of their occupants, formed part of a self-contained system that, for the most part, remained deliberately hidden. Snake-like migrant trains linking the Sul do Save to the mines slithered across relatively thinly settled coastal lowlands or the Highveld plateau or between the mountains of the Escarpment largely by night and, for the most part, well beyond the vision of white commuters or passengers. Yet, for all that, neither the SAR nor

WNLA could conceal fully the points at either end of the journey where the snake either took in or disgorged its prey – 'raw savages' on their way in or 'primitive natives' on their way out. In sluggish commercial Lourenço Marques, but more especially in modern, industrial Johannesburg, steps had to be taken to hide, as best possible, the true nature of a depressing, primitive, shameful traffic fattened partly on a diet of forced labour that neither the reactionary white public nor compliant politicians willingly acknowledged.[16]

All whites *knew* that the prosperity of the country depended on the mining industry but nobody wanted to *see* the coerced black labour that rendered the system possible and profitable. The mine compounds were designed to keep indentured black workers tied to their places of employment, but they also had the great benefit of keeping a servile labour force out of the direct line of sight of a European electorate living through its racial dream.

In an ideal world, African labour was recruited out of sight, delivered to the industrial centres invisibly, and then made to disappear into the darkness of the underground workings of the mines before being smuggled back home, also unseen, in the middle of the night. Industrial capitalism then, as now, had an aversion to too much outside scrutiny. For white South Africans, economic progress was built on the backs of a mass of black miners who were only 'temporarily' present in 'their' cities and who remained largely out of sight until they were returned to 'their' homes. African workers were part of a largely invisible labour force.

By 1926, and possibly for a decade before that, the platform at Booysens station, in central Johannesburg, had a tunnel linking it directly to 'Mzilikazi', the WNLA reception and distributing compound – the latter a structure on lower Eloff Street still partly visible behind high perimeter walling.[17] The initial, primary function of that tunnel was to safeguard the migrants by obviating the need for them to walk across a dangerously exposed section of track. But the tunnel also served some secondary functions. It would have helped curb the propensity of migrants to desert at a time when the system was still relatively reliant on forced labour, and it most certainly would have kept black recruits out of sight of the squeamish white public. In 1939, the railway management warned the WNLA that there were still too many 'batches of natives' to be seen hanging about on the main platform at Braamfontein station and that 'complaints from the [white] travelling public are bound to arise'. The WNLA needed to reassure the SAR management that its existing, commendable, practice of 'passing the natives down into the subway at the end of the platform' so that they remained as obscured as possible from the public gaze would be further encouraged.[18]

Just how squeamish organised business and members of the public could be when it came to seeing the labour entrails of the mining economy exposed was still apparent more than 20 years later. The SAR management decided that WNLA trains could not, for a time, be readily accommodated at Booysens or Braamfontein stations while the rail system was being rejigged. This meant that, for a time, the Mozambican and other miners would have to make use of the main Johannesburg Station, which catered largely for white passengers. In 1959, the resulting visual pollution – black matter out of place in a supposedly all-white world – was sufficiently serious for the General Manager of the WNLA to write to his SAR counterpart:

> With the much larger labour force now employed by the mines, the number using Johannesburg Station has considerably increased and the fact that this large number of primitive natives, often with a variety of luggage, including boxes and blankets, is to be seen on the

main platforms of the Johannesburg Station is bound to cause comment ... Natives, too, are now walking through the centre of the city to [the WNLA's] depot and this has led to adverse comment from the Johannesburg Chamber of Commerce and to a certain degree of prominence being given to the matter in the local press.[19]

For the senior office bearers in big business, the Chamber of Mines and the WNLA – English-speakers almost to a man, and most of them better educated and with a far more 'liberal' outlook than the Afrikaner Nationalist government then in the throes of refining its grand apartheid policies – several practical solutions suggested themselves. The miners could entrain or detrain from a specially designated separate platform at the central station from where they might be whisked away to the WNLA depot by bus. An alternative arrangement might be to make use of some 'isolated pocket in the railway yards'. In this instance, big business, supposedly opposed to the institutionalised racism of the government of the day, was happy enough to plough its own contours parallel to those of the state.

The Eastern Main Line and the system that encompassed it, as it pertained to the mining industry and WNLA rail transport, had been perfected over half a century. As the Chamber of Mines put it in 1930, 'years of practice have now established the general movement of recruits and repatriates with machine-like precision to and from Ressano-Garcia'.[20] The migrant labour trains, serving as an enormous transnational conveyor belt continuously moving men and adolescent boys between Mozambique and South Africa left the Sul do Save effectively bereft of agricultural labour, and of black fathers, farmers, herders, hunters, husbands, sons and lovers. WNLA commissioners scouring the region, in 1911, were struck 'by the small number of able-bodied men capable of working on the mines who had not already been there' – barely ten per cent – and arrived at the depressing conclusion that there was 'no hope of any material permanent increase being made in the number of men being recruited in these territories'. In any case, *shibalo* raids by the Portuguese had already mopped up any residue for the army or local public works.[21] It was calculated that by 1918, approximately 100 000 migrants were shuttled between Booysens and Ressano Garcia each year.[22] WNLA trains moved close on 5 000 miners to the Rand each month and close on five million black passengers to and from the mines between 1910 and 1960.[23]

From the earliest days of the system, the Chamber and the WNLA took immense pride not only in this considerable logistical achievement, but also in what they saw as the ethical treatment of all migrant labourers on the move. One of the primary functions of the WNLA, it was claimed in 1920, was 'the protection and guarantee of the fair treatment of the natives handled by it', and that it had long since enjoyed a reputation 'for fair dealing with natives'.[24] It was part of a story that was to be embroidered upon over the years until it became an integral part of what even the Chamber of Mines, adopting the language of the day, was content to designate as 'propaganda'. One test of the care supposedly lavished on the men drawn from the Sul do Save, however, is to be found in the lived experiences of black migrants during their rail journeys up and down the main line, as recorded in the words of management itself.

CHAPTER FIVE

# The Up Passage
## Mobile Incarceration

*The train bore me away, through the monstrous scenery of slag-heaps,
chimneys, piled scrap iron, foul canals, paths of cindery mud criss-crossed by
the prints of clogs.*

— GEORGE ORWELL, *THE ROAD TO WIGAN PIER*

For most of the period under consideration here, circa 1910–1955, the
WNLA ran a twice-weekly 14-coach service from Lourenço Marques up
through the border town of Ressano Garcia and then on to Booysens, in
Johannesburg. The up-trains, long numbered 804, left Ressano Garcia in
midafternoon on Wednesdays and Fridays, and were scheduled to arrive
on the Rand around midday the following day. The biweekly down-trains,
numbered 307, left midafternoon on Mondays and Thursdays so as reach
Ressano Garcia around 7 am in order to allow returning migrants to make
onward connections to Lourenço Marques and, for some, to Inhambane
(by ship). Both the 804 and 307 trains carried, on average, 450 passengers
up or down the line at a time. In one respect the entire set-up was unusual in that the WNLA, a Chamber of Mines subsidiary, was operating
a private, not-for-profit service on a publicly owned rail network. This
private-public difference extended to the control, safety and well-being of

passengers on WNLA trains being overseen by European conductors who were employed by, and reported to, the company rather than to officers of the state-owned SAR or CFLM. This arrangement contributed to underlying tensions between SAR and WNLA officials and, as already noted, helped to bedevil preferences, scheduling and travel times when it came to drawing distinctions between passenger and freight trains.

The comfort, nature and timetabling of the 804 changed significantly over four decades as the service provided for black labourers shifted slowly from transport in boxcars or open cattle and coal trucks, often with standing room only, to journeys in third-class saloons that mutated from those with single holding areas and seating to more conventional coaches with compartments leading off a corridor.[1] The changing physical configuration of trains over time obviously did much to shape the psychological and social well-being of the migrants, and this is a subject to which we will return, even if there are few surviving first-hand accounts of the hardship experienced during journeys by black migrants. The absence of personal testimony can, however, be partly overcome by asking not how the interior layout of the train influenced the well-being of the passengers but, instead, how, in a time of forced labour, the behaviour of incarcerated passengers or those with misgivings challenged the operation of the trains.

In short, *if* the transport of recruits *to* the Witwatersrand mines formed part of an entirely market-driven economy, underwritten exclusively by contracts freely entered into by black workers, it would be mirrored in the routine functioning of WNLA trains, which would be run much like any other passenger rail service albeit on a racially segregated basis. But, since we know that that was not exactly how the regional economy functioned, and that, over time, the mobilisation of black labour moved from a system in which forced labour helped prime the flow of voluntary labour to one in which rural underdevelopment and poverty eventually allowed for

71

'market forces' to direct the 'free flow' of cheap labour into an industry where real wages declined over half a century, it was always likely that this would be reflected in how recruits were controlled and treated aboard the up-train, the 804.

Between the Portuguese administration and the WNLA – the latter acting as 'a state within a state' – the mobilisation of mine labour throughout the Sul do Save was achieved through 'quasi-military' operations that continued, albeit in gradually diminishing form, until well into the 1950s. The reluctant acceptance, or second thoughts, of labourers recruited in so coerced a fashion showed in their demeanour and subsequent behaviour right from the moment they were marched out of rural WNLA recruiting camps to the station at Ressano Garcia, as well as on the 804 itself. As a leading historian of the system has noted, it was not surprising that 'mine managers naturally preferred voluntary to recruited mineworkers because the former tended to be better, more contented workers'.[2] The menace behind the recruiting process was plain to see in the contract WNLA agents were called upon to enter into with the Mozambican administration in 1918. WNLA recruiters guaranteed adult mine labourers a minimum wage of '1½ shilling per useful day's work per adult' and eightpence a day – a penny an hour – for minors, for boys. Agents were prohibited from deducting any monies from workers' wages except to cover the cost of 'transport, surveillance and feeding of the same during the journey to Ressano Garcia'.[3] Having been forced into contracts for unwanted mine work as part of a bilateral agreement in which rail traffic was traded off against human traffic, those who were coerced were then rendered liable for meeting the cost of policing en route to the border lest the more spirited among them thought of ways of exiting the system early.

At Ressano Garcia, the armed and uniformed physical force that did much to prise loose one in four of the recruits from their rural homesteads and propel them towards the border waned slightly. Inside the yard of the WNLA compound, further corralling and inspecting of recruits was overseen by men dressed in the deceptively casual fashion best suited to a subtropical climate but nevertheless wielding unfamiliar weapons – the managerial and medical eye – that seemed to hold a power all their own. A medical examination by a Portuguese doctor, 'cursory at best' suggests one expert, did the weeding, pulling out of line those with bodies already so manifestly unsuited to being sent to work in places where the sun never shone that there was no point in subjecting them to further processing.[4]

Even those passed fit, however, were frequently famished if not chronically undernourished long before they ever left Ressano Garcia. 'It was not uncommon for men to make the journey to the Rand on an empty stomach,' notes our medical historian. 'Reports of newly arrived workers gorging themselves on mealie meal in the WNLA compound attest to their deprivation during the journey.'[5] In the mine compounds themselves, for many decades before the Second World War, the poorly fed workers habitually clubbed together to supplement their meagre meat rations.[6]

In the absence of continuous 'thick' first-hand descriptions by Mozambican recruits departing from Ressano Garcia between 1902 and the mid-1950s, it is difficult to know how the rituals of departure from the WNLA compound changed over half a century, as, no doubt, they did. But here is a rare account of a departure, from as late as 1975, which is worth quoting at some length:

> At that time the new recruit still had to face many problems – for example, in Ressano Garcia they made us – the new recruits – crush the peanuts for the sauce which is strictly a woman's job! They also made us sweep the whole compound. When the time came we took

73

the train to Komatipoort. When the train stopped, it was boarded by many Swazi [Swati-speaking, South African] policemen armed with hippo [hide] whips. We, the new recruits, had no idea why these policemen boarded the train and entered our compartment, and so when they confronted us saying, 'Hei, we want to see your clothes', we were puzzled and simply gazed at them with mouths open. The old miners amongst us then said, 'To avoid any trouble, you had better let them open your luggage for inspection'! ... if you opened your food parcel they would search your luggage and if they did not find anything undesirable, they would eat your cashew, they would eat your bananas. And if you said, 'Why do you eat my provisions?', then they would take *dagga* [marijuana] out of their own pockets and say, 'You, let us go, you are under arrest because we have found this dagga on you!' He would do this because you have dared asked why he was eating your provisions. So he ate my provisions and I kept quiet. What could I have done? After they had finished searching our compartment, they came to me and said that I should go down with them, and I did that. I was made to stand in a *bicha* [line] on the platform with some of my fellow recruits. We were then ordered to go and push a cart loaded with bread and jam which we then gave out to all the miners on the train. Each miner was given half a loaf of bread and a tin of jam. We had expected that those of us who had been made to work would get an extra portion – a whole loaf of bread – but no, we only got half a loaf like all the others. I worked for nothing, indeed! This was part of the struggle in order to go to the mines! However, we ate our bread and continued with our journey until we reached Mzilikazi [WNLA compound, Booysens] where they still used the *sjambok* [whip] to organise the new recruits into *bicha*s. Many of the new recruits had caught cold on the journey ...[7]

The weapon that did most to inject slow-acting pecuniary poison into the bodies of those deemed fit to board the train for the crossing into the unknown, however, seemed the most innocuous of all – the pen. Building on the colonial fiction that none of those destined for the mines were coerced, that all contracts had been entered into freely and that all the workers would be adequately rewarded for their labour, the powers ultimately responsible for the system – the Mozambican administration and their South African counterparts – further consolidated their hold over bodies already captured by binding the migrants in debt even before they boarded the 804. The Portuguese administration, staring forever down the barrel of bankruptcy, determined to make their border bureaucracy self-financing at least and surplus-generating at best. The Chamber of Mines, intent on ensuring favourable balance sheets and dividends for its risk-averse investors spread around the developed world, saw no reason for its having to absorb the cost of running an international rail service that amounted to a third of the overall cost of labour recruitment.[8]

In what all those in the investing world around the Atlantic periphery were content to believe was a free market system underpinned by voluntary labour, it was clear how these challenges could be circumvented: by recovering as reasonable a proportion of the expense as possible and getting the most immediate short-term beneficiaries of the system – the migrants – to cover the costs of passport fees and rail fares. But how was the cash to be raised? So near-naked and penniless were the workers when they commenced the journey that they had to be provided with a 'free' issue of clothing sufficient to ensure decency and warmth, if not comfort and dignity. The cost of this rudimentary apparel was passed on to the migrants, thereby ensuring another source of income for the mine owners, a practice that Milner's post-Anglo-Boer War administration for a time debated as possibly being worthy of levying customs duty on.[9] The appeal that this business might have held for the tax-hungry government lay in the volume of 'sales'. In 1911, Lionel Phillips of the Corner House group informed his partner, RW Schumacher: 'WNLA has a fund which I believe now amounts to between £45 000 and £50 000, arising from profits on the sale of clothing to Natives.'[10]

In 1909, the basic fare of a halfpenny per passenger per mile was set for all migrants whose journey ended in Johannesburg – a fare that, by 1930, had increased by 20 per cent. When, in 1911, the WNLA decided that it was more convenient and safer for 804 passengers to detrain at Booysens rather than Braamfontein, the additional six-and-a-halfpenny levy, too, devolved upon the workers.[11] In 1914, the cost of a ticket from Ressano Garcia stood at 13 shillings and sixpence, but for most of the interwar period the rail fare and the passport fee meant that all male migrants from the Sul do Save – adults and adolescents alike – started their journey west with a hefty debt of around 25 shillings.[12] It was hardly debt bondage of the type indentured labourers experienced in most parts of the tropical world, say on plantations, since the mineworkers did receive some of their wages, in cash, with clockwork regularity. And they were spared an initial debt to the infamous company store, as was imposed on many miners elsewhere in the world. Yet, a debt of 25 shillings, recovered against wages of one shilling and sixpence per day for a shift at least eight hours in length, would take the equivalent of at least one month's hard physical labour underground to pay off – out of a 12-month contract. For the majority who had signed on voluntarily it was a huge burden; for a minority, who had been physically forced into labouring in the industry, it was unconscionable.

But, as James C Scott illuminated so brilliantly in *The Weapons of the Weak*, even the most oppressive of circumstances can excite ingenious

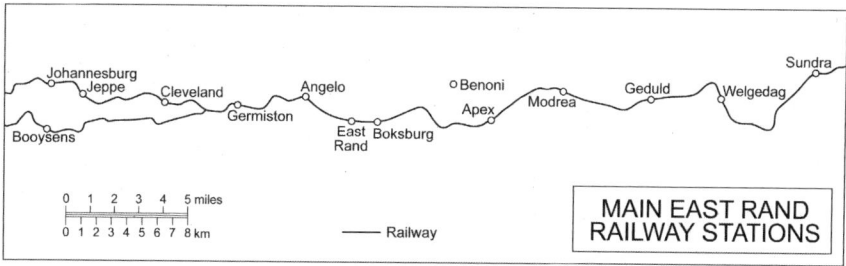

Johannesburg / Jeppe / Cleveland / Booysens / Germiston / Angelo / East Boksburg Rand / Apex / Benoni / Modrea / Geduld / Welgedag / Sundra

0 1 2 3 4 5 miles
0 1 2 3 4 5 6 7 8 km ——— Railway

**MAIN EAST RAND RAILWAY STATIONS**

forms of resistance among the seemingly powerless and vulnerable. The boldest of the migrants, unwilling to accept that their fate had been sealed by the Portuguese and their mining-house allies, or that they were condemned to dragging around the ball and chain of unwanted debt, simply inverted their predicament, thereby turning it to their advantage. Instead of seeing the 804 as a part of a conveyor belt set to dump them into industrial servitude in the coal or gold mines, they chose to view it as a service, as one merely offering them a free, albeit most uncomfortable, ride into the heart of the most promising market for unskilled labour in all of southern Africa.

The knowledge that the Witwatersrand, despite becoming increasingly racially segregated and permanently committed to a low-wage regime for all people of colour, still offered better prospects than did central or southern Portuguese East Africa was, of course, not confined to passengers put aboard the 804. Indeed, so well-known was it throughout the impoverished, *shibalo*-plagued Sul do Save that, each year, thousands of black men, women and children simply upped sticks, abandoned their homeland and walked across the nearest point on the porous South African border to seek a better life.[13] The Portuguese administration found this steady drain of one of the colony's very few resources, cheap black labour, and the erosion of the tax base through what they termed 'clandestine emigration', most alarming.[14] The full extent of that historic demographic shift and its social consequences on both sides of the border have yet to be charted fully.

On the Rand, where all black miners were routinely fingerprinted, like so many criminal suspects, two to three out of every 100 labourers absconded from the high-walled, well-policed compounds each year. Some joined the ranks of tens of thousands of 'house boys' – domestic servants – who were both better fed and better paid.[15] Many more, however, including many southern Mozambicans, did not succeed in their hopeless dash toward what seemed like a 'free' labour market. The WNLA was brutally direct and honest about their fate. Each year it reported, under the subheading 'Deserted and Sent to Gaol', the number recaptured by the state or the association's own private agents. The number of black workers convicted and sentenced for desertion from the mines was never less than 2 000 a year, and between 1902 and 1937 more than 100 000 workers were sentenced for attempting to escape from mine labour and incarcerated in prisons not unlike mine compounds.[16] How many desertions simply went unrecorded is not known.

By the late 1940s, the WNLA was using a former police detective, driving about in a custom-built Black Maria, to try and help stem the tide of deserters in yet another practice that hinted at the association's ability to act a bit like a 'state within a state'. The detective 'has an assistant and a pick-up van and is doing everything humanly possible to recover deserters', noted an official, 'but I do not think that you should look for quick results'.[17]

Those black miners recaptured and arrested for desertion were collectively labelled as 'recoveries', a purely technical term that the historically over-sensitive might be tempted to think akin to the word 'runaways' used in the old American South to characterise fleeing slaves, which, quite clearly, the Mozambicans were not. The industry nevertheless issued regular quarterly circular letters that sometimes enclosed 'a statement showing the number of deserters and recoveries'.[18] With the passage of time, mine work became steadily less attractive as other sections of the unskilled labour market expanded, and as late as 1948–1949, 22 000 deserters were sent to jail. In the history of South African industrialisation, then, the mine compounds and the prisons – or the Chamber of Mines and the state – spoke directly to one another when it came to the maintenance of a cheap black labour regime.

Confronted with the prospect of thousands of 'deserters' and illegal immigrants each year – including an unknown number making use of the 804 over the decades – the Mozambican administration and the WNLA were forced into devising formal, legal strategies, as well as hazardous informal ones, to curb the potential loss of tax revenue and workers. One strategy that impacted indirectly on the WNLA down-train dated from shortly before the First World War. In order to increase the flow of cash back into southern Mozambique and benefit local storekeepers, the Portuguese authorities persuaded the Chamber of Mines, amid predictable objections from South African storekeepers, to increase the flow of funds back into the Sul do Save by agreeing to withhold part of the black miners' wages until they returned to their homes. After an early, faltering start, the deferred-pay scheme was discontinued during war and then fully implemented in 1928. The scheme not only assisted the cash-strapped Portuguese administration and Asian traders in the Sul do Save but also helped deter potential deserters on the mines and the early, clandestine, permanent emigration of wives and children.[19]

For the WNLA, however, the problem was not simply the number of desertions from the mines themselves. That number, in any case, declined from a high point in 1912 as the newly established Union legislature, the police and the judiciary slowly tightened their grip on all Africans within

'white' urban areas. The Eastern Main Line, however, was less easily controlled, patrolled or placed under surveillance, resulting in the potential
loss of unknown numbers of forced and free migrants from the 804 while
en route to Booysens. In the first decade after the Anglo-Boer War, the
presence of a few black orderlies aboard boxcars and cattle trucks that
were simply hooked up to freight trains, along with a lack of familiarity
with the behaviour of the steam engine and its bogies, may have done
something to discourage some of the earliest migrants from deserting or
escaping either while the train was moving or when halted at various sidings. But we simply do not know the number because there was no formal
reporting on the up-journey by WNLA conductors until more dedicated
labour trains were introduced. What we do know, however, is that after
the third-class 'passenger coach' system was put in place, around 1910,
the WNLA was sufficiently alarmed by the prospect of losing some of its
'recruits' somewhere along the line for it to take special – and seemingly
wholly illegal – measures to prevent desertions.

The manner in which the up-train formed part of a system of mobile
incarceration for part-free and part-forced labour can perhaps be best
gauged from a letter written by a conductor to his WNLA manager when,
in 1913, he stood accused of being insufficiently vigilant when it came to
preventing desertions from the up-train:

> When it is considered that for a time we were using CGR [Central
> Government Railways] cattle trucks, box trucks etc. I am only
> surprised that the loss has not been greater, especially for boys for
> collieries as these natives do not wish to go there and in fact would
> refuse to go at all, if they were told that they were being sent to the
> collieries. Even when we see a boy jumping out of the train, we are
> quite powerless to stop him when the train is in motion as there is no

communication cord from coaches to engine and we can only watch the boy running away. I need not say that, in the present, as in the past, every precaution possible will be taken to prevent these desertions and it would greatly help matters if the railway had the locks put in order and the bars across the windows attended to.[20]

From 1902 onward, and for at least the next 30 years, and probably longer, the European WNLA conductors on the 804 up, denied standard-issue keys to the doors of coaches and trucks by an SAR management that feared a scandal in the event of a rail accident involving passengers trapped in sealed compartments, manufactured their own 'makeshift keys' to the doors of boxcars, coaches and other types of trucks.[21] Once clear of the mountains around Waterval Boven, and with the train set for the long, slow crawl up and across the plateau to Witbank, the conductors who, well into the 1950s, had to make sure that the number of recruits entrained at Ressano Garcia tallied with those detrained at Booysens, *locked* the coaches in a largely successful attempt to minimise the number of desertions en route to Johannesburg.[22]

By the 1930s, the WNLA was subjected to growing criticism from other colonial states in the region struggling to secure their own supplies of cheap labour for white settlers in the face of the overwhelming competition coming from Africa's Deep South. South Africa was not a signatory to the International Labour Organization's convention on forced labour and ignored most of its terms. Neighbouring governments pointed out that the WNLA's indentured workers were bound by contracts enforceable by criminal penalties. The Chamber of Mines countered with propaganda of its own, according to which 'WNLA labour was not recruited'. The association '*merely provided facilities* for voluntary workers who were determined to get to the higher wages of the Rand', and generally 'presented itself as a champion of free labour in southern Africa'.[23] But the sealed 804 told a very different story. It spoke of an unknown number of men who, however 'recruited', did *not* want to go to the detested collieries at Witbank or the gold mines some way down the track.

Locked coaches were not the only hazard that migrants had to contend with during the slow, largely joyless, steam-driven haul to Booysens. As elsewhere in southern Africa, periodic and/or seasonal congestion was an enduring feature of travel on the up-train for the better part of a half-century. The name given by Africans to most trains negotiating the SAR system was the siSwati term *bombela* – meaning 'to crowd, crush or pack' (in the post-apartheid era Nelspruit, a regional centre on the Eastern Main Line, was renamed Mbombela).[24] Particularly severe cash droughts, such as those that marked the progressive decline of the Sul do Save peasant economy for most of the decade after the First World War, almost always manifested themselves in increased WNLA recruiting, which, in turn, gave rise to periodic bouts of severe overcrowding aboard the 804. Likewise, once the planting of crops was done in Sul do Save there was a 'seasonal rush' to be recruited from January through March, which revealed itself in congestion on the hot summer up-train, and which was still in evidence as late as 1953.[25]

Overcrowding appears to have been at its most serious, reaching near-crisis proportions, between 1926 and early 1929. But, in less severe form, congestion was a feature that both pre- and postdated that crisis, and since what went up eventually had to come down, it was a feature on both the 804 and 307 trains. SAR management was of the opinion that only six men could be comfortably accommodated in the compartments of third-class coaches, amounting to approximately 50 passengers per coach. But, by October 1927, with the summer already gathering momentum for the peak seasonal temperatures in January, there were times when the 804 was already 'dangerously overcrowded', with as many as 82 migrants packed into each coach, resulting in 'natives sleeping on the floors of the compartments and along the corridors'.[26]

Ignoring its own guidelines, the SAR administration failed consistently to provide the WNLA with additional capacity. In any case, for some years it would have been of limited use since 'owing to the bridges between Komatipoort and Waterval Boven being unable to take the weight of two engines together, the train cannot be double-headed', thereby limiting the use of supplementary coaches.[27] In 1926, Portuguese border officials began to have reservations about the numbers being crammed into the 14-coach trains, but back on the Rand the appetite of mine managers for black labour was insatiable. More than a year later, the situation was not much improved, with conductors still reporting instances of seven men shoved into a compartment for journeys lasting a day and a half.[28] Throughout the 1920s there were instances harking back to the years following the Anglo-Boer War when cattle and coal trucks were employed to cope with unexpected volumes of migrant traffic. As the Chamber of Mines had to point out to the General Manager of the SAR, which was always more freight- than passenger-minded, 'besides the excessive heat on this type of coach [sic], there is no lavatory accommodation or fresh drinking water available'.[29] Even the WNLA baulked when, on a day in January 1929, there was only one coach available for 108 passengers and the station inspector wanted to bundle 50 repatriates into a cattle truck bound for a sweltering Ressano Garcia.[30] Two decades later, admittedly on rare occasions, some Sul do Save migrants were still standing for the entire journey from the border to Booysens.[31]

Crawling across the windless Highveld plateau beneath a midsummer sun, whether in trucks designed to convey cattle or in congested modern third-class 'saloons', meant that overheating, adequate supplies of potable water and a lack of ablution facilities were enduring problems. There were, WNLA officials claimed, 'many cases' in which migrants, presumably seeking relief from the heat and overcrowded conditions, took to 'sitting in open doorways' with their legs protruding beyond the coach or truck. The sequel was largely predictable. While passing over a culvert, 'his legs came in contact with bridge' and 'he was drawn out of the train'. And because, as we have already noted, WNLA trains were not fitted with an alarm system until at least the end of the First World War, the conductor who witnessed the incident recounted above 'was unfortunately powerless to help'.[32]

As already noted, the British administration that had fought a war to help advance civilisation north of the Vaal River, was initially not very enthusiastic about the idea of providing third-class passengers with drinking water.[33] Matters improved slowly after Union, but throughout the 1920s and 1930s there were complaints about faulty water pumps on passenger coaches that remained unrepaired for weeks on end, while at Komatipoort there was a time when the water quality was officially condemned.[34] Inadequate water supplies had a knock-on effect, causing problems with malfunctioning ablution facilities. As one stationmaster noted, even in cases where black passengers could enter or leave 'coaches' freely, 'Much offence was caused by natives easing themselves on the line, in full view of white passengers.'[35] What happened within the confines of enclosed or locked coaches over 18 hours can be imagined. In 1923, two experienced WNLA conductors resisted, in writing, the idea of detaining returning Mozambican migrants for an additional 50 minutes at Witbank so as to accommodate a proposed change in the timetable because 'apart from the sanitary factor, a delay of this duration makes the natives restless'.[36]

Acute discomfort occasioned by absent or inadequate basic facilities, along with their attendant dangers, was dragged out for what, to the migrants, must have seemed like eternity.[37] Slow progress was not only a function of the habitual preference given to freight over passenger trains, but was also caused by SAR officials having to employ engines lacking sufficient steam power to haul the 804 up and onto the plateau with ease. Through the 1920s, the most frequently used locomotives were the Class 14A. 'This engine,' complained an irritated conductor on the last leg of his homeward-bound journey in 1926, is 'supposed to pull eleven hundred tons' yet it cannot 'pull 400 tons from Germiston to Booysens without getting stuck en route for want of steam.'[38] Four years later, a colleague was moved to complain: 'We left Waterval Boven at 1.00 am and arrived at Dalmanutha (18 miles away) at 5.00 am, in the interim being subjected to severe buffeting caused entirely by the engine being incapable of handling the load.'[39] The temper of hungry and thirsty westbound recruits might not have been improved had they known that Dalmanutha – named after the coastal town identified by Mark in the Bible – was Jesus and his dis-ciples' destination after the masses had been fed with the loaves and the fishes. Miracles were at a premium on the 804, where only half-loaves and tins of jam were in evidence and tins of pilchards had to be purchased independently by the migrants.

There were, of course, reasons other than being personally inconven-ienced by long shifts that pushed WNLA conductors into issuing written complaints about inadequate Class 14A locomotive power for the haul up onto the Highveld, or about the slow movement of the 804 through the congested eastern sector of the industrialising Witwatersrand. 'This has to be experienced to be believed,' wrote a conductor of a migrant train trying to negotiate a 'stiff bank' near Dalmanutha, and it 'means a spell of agony for the passengers.'[40]

But, as the saying has it, it is an ill wind that blows nobody any good. In 1920, the up-train was noted for taking close on two and a half hours to cover barely 27 kilometres on the Rand.[41] 'The train travels so slowly in places,' it was pointed out 36 months later, 'that the boys can step off with impunity.' 'Moreover, delays of the kind like we [have had] at East Rand, are direct inducements for boys to desert.' 'If the train was run properly between Geduld and Booysens we would probably have fewer desertions.'[42] During the Second World War, a five-kilometre stretch between Delmore and Germiston became notorious for giving priority to as many as 20 passenger trains before allowing the 804 to proceed. It made the Mozambicans 'very restless', who then, 'complaining of hunger, ask to be allowed to walk the remaining distance' – a request that, for all the usual reasons, starting with desertion and ending with their personal safety, obviously could not be entertained.[43]

In winter, a lack of steam power and painfully slow progress at elevations of more than 1 700 metres above sea level ushered in a different sort of terror between Waterval Boven and Witbank, where the temperature at night frequently hovered around freezing point. Travelling in boxcars and cattle trucks or coal trucks on either the 804 up or the 307 down – reasonably consistently before 1910, and then occasionally thereafter for two more decades – would have contributed directly to the appalling death rates from pneumonia over many years. During the interwar period, engine drivers – often saddled with ageing locomotives lacking the capacity to take on long inclines more comfortably – were loath to forego much-needed boiler pressure by releasing steam for the purpose of merely heating coaches. A conductor, reporting from Komatipoort in midwinter 1932, knowing what lay beyond the mountains at Waterval Onder, had this to say: 'I have to report that in only about five trips out of ten is the steam on at all, either owing to the pipes between the coaches not being coupled up, or the valves being open, or there being no steam-pipe connection to the engine.'[44] Negligence, stupidity and possibly even spite on

86

the part of lowly white operatives notorious for their unwillingness to see the 'Kaffir Mail' train as part of a conventional passenger service meant that the train was often left grossly underheated, if it was heated at all.[45]

Sadly, only a few surviving fragments drawn from the lyrics of one of the many versions of *'Shosholoza'* (such as the rhythmic chanting of *'puma ghazi, puma ghazi'*, meaning 'to sweat heavily', as when engaging collectively in heavy manual labour) are all that we have to hint at how the Sul do Save migrants relived such intensely corporeal experiences aboard the 804 up or 307 down. So, in the absence of first-hand testimony as to how the challenges of the train and what lay beyond were seen by passengers, the best we can do is to track their behaviour in the hope that it casts light on how such experiences were anticipated, endured and internalised not only by the migrants, but also by their family, friends and co-workers.

Henri Junod, Cartesian to his very core, a Swiss missionary and the pre-eminent ethnographer of the Shangaan-Thonga in this era, had a profound interest not only in the past but also in how traditions passed down to the men and women of the Sul do Save articulated with the realities of a new and rapidly changing social environment. He respected and understood the nature of the demands imposed upon young men at initiation lodges but was puzzled by the longevity or relevance of the ordeal, and reluctantly came to the conclusion that 'these rites as a whole have very little worth and are useless in the new economy of South Africa'. Yet, his list of six 'main trials' in preparing novices for adult life reads, in part, like a checklist of tests to be faced by any young man or adolescent bound for the 804, the 307 and the Rand mines: 'blows, cold, thirst, unsavoury food, punishment and death'.[46]

The train may have embodied the essence of 20th-century modernity for Europeans, but when young black males were cast as, or treated as if they were, animals or mere commodities, it harked back to a premodern era and called anew for the mastery of the most rudimentary of survival skills. Colonisation by military conquest entails an element of moral regression on the part of the invader, and thereafter consistently complicates any simple-minded notions of what might qualify as 'progress'. In a situation in which two colonial powers foregoing most ethical considerations had, without consulting the party principally affected, agreed to the wholesale appropriation of African labour at exploitative rates in exchange for rail traffic to a port city that otherwise might not stand up to open capitalist competition, the past and the present were not that easily divided. Supposedly backward-looking 'tradition' acquired new and longer-lasting relevance precisely because the colonising victors employed 'modern' technology in ways that denied common notions of humanity and exercised a stunted morality. When it came to the functioning of the 804 and 307, it is difficult to know who the 'raw savages' were – the initiators and operators of the supposedly 'modern' rail system or their more tradition-bound migrant captives?

The crude and dangerous nature of migrants' corporeal experiences aboard the trains and on the mines added significance to 'rites of departure' that predated the inception of the Eastern Main Line. Diviners dispensed an assortment of protective medicines to protect 'the migrant from accidents, disease, and bewitchment'. More important still, these rituals ensured that 'if the young man died far from home, his spirit would enter in the world of the ancestors' – a consideration already hinted at and one that we will explore further.[47] The rites may also, however, have offered recruits a timely reminder that not all hazards were physical in nature, and that the train was capable of unleashing some profound psychological disturbances.

In 19th-century Europe, many writers, including doctors coming to terms with the exceptional growth in rail travel, drew attention to accompanying symptoms of physical or mental stress – a condition that a modern-day historian still records as 'railway shock'.[48] Passengers who felt trapped in unfamiliar confined spaces, including some who presumably were reasonably well-educated and literate, feared accidents and the trauma of possible derailment. Effectively incarcerated for as long as the train was moving, still others became unnerved in compartments where the bodies of strangers were brought into uncomfortable, or unnatural, proximity to them.

More disturbing still was the way in which the train was capable of destabilising the senses as time, space and vista became concatenated, giving rise to greater or lesser moments of confusion and disorientation. 'When an early film of a train pulling into a station [in Europe] was shown,' one modern analyst observes, 'its audience, new to both trains and moving images, reacted with panic.'[49] Durkheim had already suggested that society's space-time perceptions were a function of social rhythm and territory. That being so, one might imagine, the swifter and starker the transition between a rural universe pulsed by the seasons and an urban dispensation governed by the clock and industrial discipline, the greater the potential for disorientation as to time and space – the very bedrock of an individual's mental well-being.[50] Any such problem might be further exacerbated by the effect of fatigue on a person's circadian rhythm.

There is, alas, no way of comparing qualitatively the conventional 19th-century European exposure to 'railway shock' with that of African peasant-proletarians in southern Africa during the early 20th century. We can hum the tune but now lack the original, all-revealing lyrics. Perhaps all that can be safely suggested is that, for first-time migrants sucked out of the Sul do Save and then pumped into the compounds and underground workings of the Rand mines, the experience may have been no less unnerving than that undergone by Europeans in peacetime and was probably far more so. In addition to the physical hardships already recounted, migrants underwent the personal humiliation and shame of having to defecate or urinate in overcrowded, locked coaches moving for a good part of the time at a snail's pace through unfamiliar mountainous terrain by night. Taken in the round, the journey was often traumatic and, for many, the start of a life-changing experience.

Everything we know about the migrants' preparation for the trip, the physical and psychological discomfort experienced aboard the 804, and the determination of some to escape the clutches of the WNLA would suggest that, for many, the trip west was part of an extended ordeal – an ordeal drawn out for a further 12 to 18 months once on the mines. From at least 1911, and for decades before that, it was known that a tour of duty on the Kimberley or Rand mines placed the 'stamp of manhood' on the men or boys of the Sul do Save and, in that sense, was either a precursor to undergoing the trials of circumcision, or supplemented them if already experienced.[51] But the point being made here is that the train journey up to the mines *itself* formed an integral part of the threshold – or liminal – experience that labour migrants underwent. It cannot and should not be fully separated from either traditional cultural practices that preceded it or from the identity changes that followed in the mine compounds or while working underground. A boy who had successfully negotiated the traditional Thonga passage to manhood was said to have 'crossed (*wela*, like a boat across a river)' and it hardly stretches the imagination to see how train journeys, too, might have come to symbolise a 'crossing' of sorts.[52] The 804 ordeal marked a moving out of traditional structures, either through coercion or voluntarily, and imposed a new collective status on recruits in an unknown social and working environment.

# PART III

# CHAPTER SIX

# Masculinity and Madness

## The Mental and Physical Disposition of Exiting Migrants

*The restlessness and the longing, like the longing that is in the whistle of a faraway train. Except that the longing isn't really in the whistle – it is in you.*
— MEINDERT DEJONG

Just as the functioning of the 804 up-train and the behaviour of many of its reluctant passengers cannot be fully understood without knowing how the Eastern Main Line fed off a centuries-old Mozambican territory with a history of slavery, and indentured and *shibalo* labour, so the operation of the 307 down-train with its thousands of human discards cannot be assessed fully without first situating it within the wider context of the nature of the Chamber of Mines' indirect hold over migrant workers. And while the mining industry's major – one might even suggest its sole – concern was to optimise the use of black muscle as a source of machine-like, ultra-cheap labour, we need to pause and explore briefly the crucial body-mind connection.

The dismantling of the thousands of individual personalities of the migrants, whether press-ganged or voluntary, and the systematic remoulding of separate personas into the collectivised 'batches' better suited to the needs of the mining industry, commenced in the WNLA's remote recruitment camps deep in the rural Sul do Save. At Ressano Garcia, the

93

issuing of metal dog tags, worn around the neck and bearing a Portuguese passport number, advanced this process of objectification, albeit as a secondary rather than a primary objective. Transportation in crowded boxcars, cattle or coal trucks, and coaches while being perceived by racist railway managers as freight rather than as passengers only lent itself to further collectivisation, commodification and dehumanisation.

At Booysens, inside Mzilikazi, the WNLA's reception, holding and labour redistribution compound in lower Eloff Street, albeit under a different name, additional bureaucratic processing and medical screening made it clear that the mining industry's needs were predicated on the primacy of bodily, rather than mental, health.[1] As the WNLA's manager in Johannesburg complained to his counterpart in Lourenço Marques in 1922:

> Amongst the natives received within the last three weeks are a considerable proportion who are so weedy and undersized that it is almost impossible for us to place them. You will realise that at a time when there is an abundant supply of labour mines become more particular as to the type of native whom they will accept. At the present time only good hammer boys are placeable.
>
> I appreciate, that as the natives that you are now sending come up mainly from the famine districts their physique is bound to be sub-normal, but it is quite useless to send natives so much below the mines requirements as those I am referring to. Further, it would really be better for you to detain the natives at Maxixe [Xai-Xai] and feed them there than to forward them to us here for feeding.[2]

On mine properties, the rudimentary accommodation of migrants in crowded – often overcrowded – barracks and concrete-bunk-type

94

housing, hidden behind high walls, with a single, well-policed entrance that not only excluded any outsiders but also all women, proved to be an enduring physical feature of the industry throughout the first half of the 20th century.[3] As Nelson Rolihlahla Mandela, who at 23 worked briefly at Crown Mines as a black compound policeman for a few months in 1941, later recalled it:

> I was given a uniform, a new pair of boots, a helmet, a flashlight, a whistle and a knobkerrie, which is a long wooden stick with a heavy ball of wood at one end. The job was a simply one: I waited at the compound entrance next to the sign that read 'BEWARE: NATIVES CROSSING HERE', and checked the credentials of all those entering and leaving.[4]

Mandela's efforts as a compound policeman during the formative years of his political awakening were sufficiently conscientious for him to remain on good terms with the managers of the mining industry – black and white alike – for several months after he left Crown Mines. In 1942, in order to save money while working as a junior in a law firm, Mandela was given the benefit of living rent-free in the WNLA city compound – an unusual privilege for a non-miner, but he was also a man drawn directly from the Thembu royal house then actively supplying black mine labourers from the Transkei.[5]

Always the jewel in the crown for the Chamber of Mines when it came to the control of African labour and the denial of the right to black workers to organise collective industrial action, the perpetuation and maintenance of the integrity of the compound system – as distinct from attempts by the state to mitigate conditions behind compound walls – was never negotiable. The compound system was tolerated, successively, by the Kruger government, the post-war Milner administration bent on modernisation and reconstruction, and every single government thereafter, from Louis Botha at the time of Union right through to the advent of the Afrikaner nationalist administration under DF Malan in 1948.

The mine compounds were part of a factory-like system that helped entrench notions of inferiority and servility in black populations throughout the region. Together with the ghetto-like urban 'locations' and the 'native reserves', the mine compounds formed a vital cog in the articulated geospatial machinery that pumped economic life into the emerging systems of first segregation and later apartheid. But not even the delegates at the inaugural conference of the South African Native National Congress (the forerunner of the African National Congress) in 1912 sensed how the mine compounds already had, for close on two decades, undermined both the physical *and* the mental well-being of African miners and that the two elements were closely intertwined. Thus, while the conference asked for workmen's compensation for death or injury to be extended to African miners, it declared that it was not, in principle, opposed to the building of compounds provided they were managed by 'competent and sympathetic people'.[6] As with the WNLA, it was the black body rather than the black mind and personality that mattered most, and the disposition of the jailers rather than the jail that posed that greatest menace to African men.

In 1930, the old but still profitable Simmer & Jack Mine – hardly exceptional – housed 40 per cent more black miners than was legally permissible, and as late as 1964, one in five compounds in use on the Witwatersrand had been constructed before the First World War, with many manifesting the attendant, decades-deep problems of lighting, space and ventilation.[7] Nor were conditions underground, where high accident rates were common, significantly better. Basic health and physical working conditions remained poor, and the shift system and work rhythms imposed on black recruits were backed up by written and unwritten codes of conduct and discipline that were often enforced by brutal, arbitrarily inflicted assaults by white miners on African workers that seldom ended up before the courts.[8]

The casting of adult black males as 'East Coast Boys' by mine management and white miners might have helped trivialise illegal assaults that humiliated black victims but they did, albeit very grudgingly, acknowledge the humanity of those attacked in a way that denied supposed 'minors' the right of self-defence. By denying black men the right to insist on the use of their proper, given names, or refusing to give them more conventional nicknames that acknowledged their status as human beings, white miners added to the processes of dehumanisation and objectification noted above. The propensity of white South Africans to give black miners names of things – Pencil, Stick, Sixpence, Sjambok or Matches – just added to the humiliation.[9]

Taken in the round, then, the black migrants' experience of recruitment, transportation, housing or work on the mines was not generally conducive to cultivating a sense of inner well-being or self-confidence. Some young miners, no doubt, managed perfectly well, perhaps even enjoyed – up to a point – overcoming the extreme personal challenges that came in a racist environment characterised by hyper-masculinity. Unfortunately, few 20th-century southern African researchers bothered to question deserters, or returning Mozambican miners, about how their experiences on the Witwatersrand mines contributed to their character or personality at a formative stage in their life cycle. It would, however, be surprising if, for some at least, life in a mine compound – an experience that in some ways paralleled confinement in an internment or prisoner-of-war camp or a prison – did not impair their dignity, alter their self-image or undermine their self-confidence. Working on the gold mines meant having to cope with unremitting mental and physical demands. Our understanding of these painful issues remains rudimentary.

In South Africa, as elsewhere in the developed world during the first half of the 20th century, the gold mines were characterised by assaults on mind and body alike, with consequences capable of flowing in both directions. In America and the United Kingdom, medical science was slow to come to terms with the ways in which abnormal psychological stresses encountered in the collieries in Scotland or the United States might impact on the health of mineworkers.[10] On the Witwatersrand, it would appear that the only people who were slower to recognise the importance of the links between the mental health of black migrants and their physical well-being were the doctors at the time and, later, medical historians. These links are only now beginning to be explored, but much more research needs to be done before we can come to a more precise understanding of the experience of black Mozambicans on the gold mines. Until we have more professional studies, many of the stresses arising from accidents and deaths will have to be inferred indirectly from the behaviour of the migrant miners.

Throughout history, new mining towns – where men often outnumber women – have been associated with masculine excess and the release of pent-up tension though aggressive behaviour, fighting, heavy drinking, compulsive gambling or the purchase of sexual services. Over time, such frontier-like indulgence tends to recede, if not disappear, as the composition of towns in terms of age, gender and sex ratios comes to assume a more familiar social profile. On the Witwatersrand, however, the mining industry's insistence on the compound system as an instrument for the control of most black migrant workers ensured the perpetuation of a family-deprived, high-stress and frequently loveless environment for half a century. There was no 'normal' industrialisation in what was an abnormal society.

The three-way link between South African mines, the Sul do Save and excessive alcohol consumption dates back to the discovery of diamonds in the late 1860s. The Thonga were among the first long-distance migrants to seek work on the Kimberley mines, and back in Mozambique, Indian storekeepers and Portuguese colonists were quick to note an uptick in disposable cash income in Lourenço Marques and the rural hinterland. In metropolitan Portugal, the news came as a godsend for a country that lacked a broad-based manufacturing sector and struggled to sell its wine surplus in increasingly competitive and discerning European capitals. The already rapidly developing market in southern Mozambique positively boomed once gold was discovered on the Witwatersrand in the mid-1880s. Between 1889 and 1898, wine imported into Angola from Portugal increased by 17 per cent, but over the same period the amount of *Vinho Para O Preto*, 'Wine for Blacks', imported into Mozambique – with the overwhelming bulk of it adulterated and destined for the Sul do Save – increased by a whopping 1 123 per cent. A growing dependence on alcohol and outright alcoholism became an increasingly prominent and depressing part of black social life on the mines and in southern Mozambique from the 1870s onwards and, in the latter case, remained a prominent feature until well into the early 1920s.[11]

Predictably, an increased physiological need for alcohol – as opposed to developing a consumer-centred taste for wine – eventually translated itself into a preference for spirits. For the men of the Sul do Save, a marked shift in taste towards strong spirits in the early to mid-1890s coincided with a move from labouring on outcrop mines to working at far greater depths in the more dangerous and stressful deep-level mines. The mine owners, sensing a chance to recoup some of what was being spent on black wages and an opportunity for improving the skills and productivity levels of the workers by reducing their personal savings and prolonging the time that migrants spent on the Rand, initially *opposed* prohibition

and invested heavily in a local distillery – Eerste Fabrieken – supplied with
fruit and grain by the otherwise God-fearing Boer farmers. By 1895, there
were seven other distilling companies operating along the border around
Ressano Garcia.[12]

The idea behind the unrestricted sale of alcohol to off-duty black miners, however, had a few inherently contradictory outcomes, and the logic
behind the scheme unravelled when large-scale drunkenness led to a
reduction in the size of the pool of sober and productive black workers
available to the industry on any one day. The official estimate was that 15
per cent of the workforce was incapacitated at any time, but unofficial
estimates put the figure as high as 25 per cent – one in four – a number
that tallies, more or less, with the number of increasingly opium-addicted,
incapacitated Chinese miners who worked on the same mines between
1902 and 1910. The history of the first 20 years of the Rand mines – arguably the most dangerous decades ever experienced in the industry in terms
of death rates – was characterised by foreign workforces that were significantly drug-addicted. It was, and remains, almost impossible to uncouple
minds from bodies in the industry.

As the mine owners came to realise during the 1890s, any gains to be
made indirectly, via shares in the distillery in Pretoria, from the sale of
alcohol to the numerically predominant migrants from the Sul do Save
were outweighed by what was lost through the loss of a sober and productive workforce. So they became champions of prohibition, which duly
became the law in 1896. This was followed, in 1900, by a ban on the production and distillation of spirits within Mozambique itself, but not on
the sale of wine.[13]

For the better part of 30 years, from around 1890 up to about 1920, with
the odd interruption such as that occasioned by the Anglo-Boer War, the
Chamber of Mines, mine managers and WNLA agents helped sustain,

and unashamedly made use of, the pernicious primary link that the Portuguese had forged between alcohol, the Sul do Save and black miners on the Witwatersrand to control the mine owners' favoured pool of cheap and captive labour in southern Mozambique. When the production of cheap spirits destined for the African market was progressively dismantled on both sides of the border, between 1896 and 1902, the Chamber of Mines, reluctant to forego the link entirely, persisted with a low-alcohol 'kaffir beer' ration in mine compounds. As late as 1917, WNLA agents were still making use of dance and wine parties in remote Sul do Save camps to groom potential recruits, and until at least the Second World War seriously or terminally ill migrants on the 307 down-train were provided with brandy to ease their pain – a concession never granted legally to other black workers.[14] In several respects, then, when it came to the well-being of black industrial workers, as opposed to that of white miners, the mining industry in South Africa and Mozambique was either beyond the immediate reach of the law or the beneficiary of strategic exemptions and a studious, state-tolerated blindness.

As elsewhere in the world, Mozambican miners probably used alcohol to help cope with work-induced stress, or simply as a reasonably familiar lubricant to sustain social relationships in an overcrowded, potentially volatile, all-male and generally oppressive environment. For much the same set of reasons, one imagines, recruits used the up-train to smuggle in large quantities of cannabis for personal use in the compounds – yet another part of an age-old rural tradition bent to help meet the stress imposed by a brutal form of modern industrialisation.[15] In the case of alcohol, however, the point is that nowhere else in the civilised world was the link between the producers of beer, wine or spirits and a mining industry spread over two neighbouring countries as closely and consciously fostered, or for so long a period, as it was on the Witwatersrand.

But, as importantly for our purposes here, we will never know the full cost in mental or physical terms that alcohol dependence or heavy use exacted directly on the men, and indirectly on the women and children, of the Sul do Save. Nor is it always helpful to distinguish so clearly between the mental and physical manifestations of addiction, because they were often closely intertwined. It is, for example, now well-established that there is often a link between alcohol abuse and what today is regarded as post-traumatic stress disorder. In industrialising South Africa, the mining industry was party to a sustained attack on the characters, minds and bodies of African men.

If the use, or excessive use, of alcohol and cannabis was, in part, an attempt by Mozambican migrants to ameliorate some of the harsher consequences of alienation and stress within the self-contained world of compound life, then so too was the subculture of homosexuality on the mines. Some sons of black miners, mere boys, it was alleged by the British Consul in Lourenço Marques in 1915, were subjected to rape by 'foreign nationals' – WNLA recruiters – in the Sul do Save even before they became of working age.[16] Many other affectionate and intimate couplings that brought together influential, powerful, senior males with more disposable income than most and numerous younger men, ranging in age from 15 to 25, might have evolved as genuine, long-lasting and reciprocally loving relationships in any urban or rural social setting. Others, however, might have been innovative and temporary responses – and no less emotionally rewarding for all that – to a marked absence of care and compassion under otherwise brutal compound conditions.[17]

Yet, when seen within the wider context of the migrant miners' experience in the first half of the 20th century, it would seem that the majority of those relationships fell squarely into the latter category and were probably defined more by the situation than they ever were by free choice and reciprocal love.[18] And, beyond that, there clearly existed an even darker side to such sexuality that was a more or less permanent feature of compound life on the Witwatersrand and was never far from violence.

In a mineworkers' song still heard in the rural parts of the Sul do Save in the 1960s and 1970s, *all* the men who went to the Witwatersrand were referred to as 'Rosa'. When asked why his countrymen had been given what was so obviously a female name, a field worker recorded his view that 'the Mozambican miner becomes "Rosa", the woman, on arrival on the mines because he is more or less inevitably subjected to homosexual rape there'.[19]

But regardless of whether such relationships were freely entered into, solely as a result of sexual preference, or were more in the nature of opportunistic and temporary liaisons entered into by males of greater or lesser vulnerability in an otherwise emotionally stunted environment, none were immune to the conflicts, resentments and jealousies that come with most human relationships. For that reason, they too could exact a severe mental toll on black workers who were unable to freely negotiate the wider social world.

In one of the few historical studies of Mozambique we have that outlines the nature of these complex relationships – couplings that mine managers often approved of because they could helped prolong the work cycle and the acquisition of skills – the potential for emotional damage is clearly outlined. 'The initial stage of the homosexual relationship marked a traumatic imposition of male authority,' suggests a historian of the Sul do Save, and some informants recalled 'deep emotions that at times led to suicides, murders and beatings if the boy left his husband for another man'.[20]

The 'moral' and 'social' problems of mine marriages and alcohol and cannabis abuse – but seldom to the same extent the underlying contributory

# THE NIGHT TRAINS

causes that an absence of family life or appropriate working class housing induced – aroused the interest of both church and state at various junctures during the first two decades of gold mining on the Witwatersrand. When the family was conveniently compartmentalised and uncoupled from the realities of compound life in that way, there was even less reason to link the question of the mental health of black migrant workers – small in number and even less visible to outsiders – back to the same cluster of issues. Yet, for all that, and as we shall note in due course, WNLA officials *knew*, from the turn of the century, that the pressures and stresses of mining left an unknown number of men withdrawn, deranged or extremely violent.

As early as 1903 – as we will explore in Chapter Seven – the WNLA was aware of the need to make special provision aboard the coaches of its 307 trains for the shipment home of those Mozambican miners who had gone insane while working on the mines. Only further in-depth research in Johannesburg or Maputo and the Sul do Save will hint at the extent and nature of the mental onslaught experienced by black miners who went to the Witwatersrand during the first half of the 20th century. In the absence of such quantitative or qualitative data, the best we can do, for the moment, is to note, in passing, how other migrants who were exposed to broadly similar mining experiences in central or southern Africa at the time became mentally unstable, even if the pickings from comparative studies, too, are also wafer-thin when it comes to focused evidence.

A single surviving file detailing the medical history of 654 Chinese miners repatriated from the Rand compounds in 1905 after only a year's work underground offers us comparative, albeit tangential, evidence of how severe was the mental onslaught on workers of colour. All the workers had undergone repeated medical examinations prior to their departure from

China, during the voyage by ship to South Africa and after their arrival on the Witwatersrand, and 99 per cent of them were between 20 and 39 years of age.

By the time the mining industry repatriated them, after barely 12 months' underground labour and compound incarceration, their disorders were said to fall into two broad categories – the bodily and the social. Not surprisingly, just over 18 per cent of the repatriates were suffering from respiratory diseases such as silicosis and tuberculosis; 12 per cent had developed musculoskeletal deficiencies; another 10 per cent had been either seriously injured or blinded while working; while yet another 6 per cent had become infected with syphilis. Crudely stated, around 46 per cent, almost half of those repatriated, had been reduced to bodily wrecks – many, perhaps most, never to recover.

Such a heavy physical toll on bodies was perhaps to be expected from labouring underground in mines being extended to ever-greater depths amid a sustained managerial drive to cut costs. What is more revealing, however, is that of the 654 sent back to China, 12 per cent were addicted to opium (a percentage that had risen markedly during their stay on the Rand) and an extraordinary additional 22 per cent (by far the biggest category of 'rejects' being returned to Asia) were said to be suffering from social maladies of one kind or another – close to one in three. And of those 143 'social' wrecks, no fewer than 83 were said to be suffering from 'senile decay' or were deemed to have become 'mentally deficient' or 'half-witted'. Another 23 were considered to be engulfed by a 'weak-minded mania', 'violent' or just plain 'insane'. Even when one considers the possibility that at least some of those suffering from these 'social disorders' might have faked or exaggerated their symptoms, one would have to ask what drove men to take such desperate measures.[21]

Two decades later, in the early 1930s, of 142 'invalids' repatriated to colonial Malawi from Southern Rhodesian (Zimbabwean) coal and gold mines, 12 were characterised as being 'unstable or insane'. In the 1950s, among the Bemba of colonial Zambia, 'the cry often heard in the village is – "He went off to the mines to work and came back mad."'[22] We also know that in South Africa, by the late 1930s, some of the Rand mine hospitals were already employing 'male mental nurses' because, in 1940, the all-powerful Gold Producers' Committee of the Chamber of Mines set about codifying the conditions of service and specifying the qualifications for such staff members.[23] By 1950, and perhaps for a decade before that, one or two heavily sedated black Mozambican miners deemed to be potentially dangerous were being sent back to Ressano Garcia on the down-train every week. They were placed in straitjackets once the sedation wore off, before being passed on to a hospital in Lourenço Marques.[24] And Mozambicans who worked in the mines in the 1970s recall that some who returned to the Sul do Save were said to be *xilutse lutse* – 'mad' in Tsonga.[25]

And, from an isolated case recorded by a couple of sharp observers visiting a mental hospital at Mafeking in 1973, we know also that the manifestation of madness in a migrant could be linked directly to the SAR-WNLA train – the very same open-mouthed mechanical snake with a near insatiable appetite for raw black labour that swallowed men whole in the Sul do Save. Although the researchers' observations come from the far west of the South African Highveld rather than southern Mozambique, they nevertheless cast a bright, reflected light on the 307 train, and their vivid description of the patient they encountered in Mafeking is worth quoting at some length:

His crazy clothes spoke the language of his obsession. His boots, standard issue for mineworkers, were topped by intricately knitted leggings, the painstaking product of many unravelled orange sacks. He wore a cloak and bishop's mitre, fashioned from black plastic garbage bags. Across his chest was stretched a brilliantly striped sash, on which were stitched three letters, SAR. For his white attendants, these were the most obvious signs of his delusion, although they also noted that he 'heard' things. The other patients, however, regarded him as an inspired healer, sent to them in their affliction. SAR was his church, and he was its sole embodiment. The letters stood for South African Railways, alongside whose track the hospital lay. In fact, at the very moment we encountered him, the night train for Johannesburg rattled by with its daily cargo of migrants. Later, as we puzzled to decipher his message, we kept returning, as he did, to SAR. It was a message that spoke directly to his fellow inmates – and also to the black paramedical staff. For, in this world of peasant-proletarians, the railway forged a tangible link between rural and urban life, hitching together the dissonant worlds of the country and the city.[26]

Poignant and painful as the calling of Bishop SAR was, and needy as his delusional flock might have been, in terms of sheer numbers their inner anguish had long since been eclipsed by the physical assault inflicted on countless other black bodies for more than half a century. The needs of northern-hemisphere bankers, financiers and speculators 'risking' their hard-earned capital to recover gold from thousands of feet beneath the surface had unleashed the equivalent of a full-scale war, one fought at close quarters in exceedingly cramped quarters on the most unequal of terms. As often happens in street-by-street engagements – or, in this case, rock-by-rock, stope-by-stope fighting – the war produced tens of thousands of casualties and fatalities. And, as in many a bloody conflict, there was a train to evacuate the living dead and those who had been fatally wounded in the struggle for an ostensibly noble cause that was under-stood only imperfectly by the idle, the poor, the vulnerable or the weak. In the Sul do Save, where the war had 'smashed the warriors' – or, as some chose to see them, 'circus animals' or 'raw savages' – the train was omni-present in the consciousness of the people. There was the up-train that took 'recruits' away, into a war for the civilising of their minds through industrial labour, and then there was the down-train, which brought back what was left of the bodies and minds of the veterans. As the Transvaal Government Mining Engineer noted in 1901, 'Mining is more dangerous than war, but the miner gets no medals.'[27]

Besides individual and collective mine accidents, which accounted for thousands of black lives, three other serial killers stalked mine compounds insistently during the first two decades of the 20th century and on, with sharply declining intensity, into the 1950s: pneumonia, silicosis and tuber-culosis (TB).[28] Each had its own aetiology. In outrageously broad terms, pneumonia was an occupational disease in the sense that it often arose from the failure to provide adequate change-room facilities for miners making the transition from working in moist and warm underground operations to recovering in the structurally deficient compound condi-tions.[29] Silicosis, 'a disease characterised by fibrosis of the lungs due to the inhalation of silicious dust', was the occupational disease par excellence in the mining industry.[30] Tuberculosis, caused by the 'inhalation of airborne droplets containing tubercle bacilli emitted by persons with active TB', was a disease that could, and had, occurred in many different countries and settings but thrived in overcrowded mine compounds where, it was conveniently and long suggested, Africans had a peculiar and supposedly racial susceptibility to the infection.[31] Insofar as congested living condi-tions were a cause of TB, it too was – in part – an occupational disease for black miners almost exclusively.

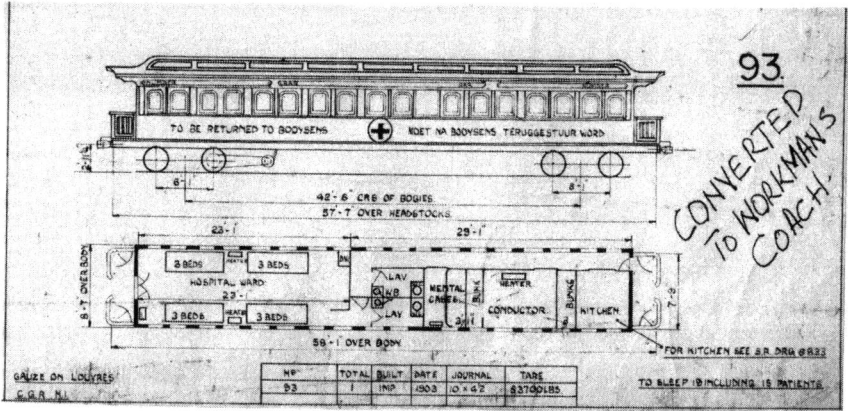

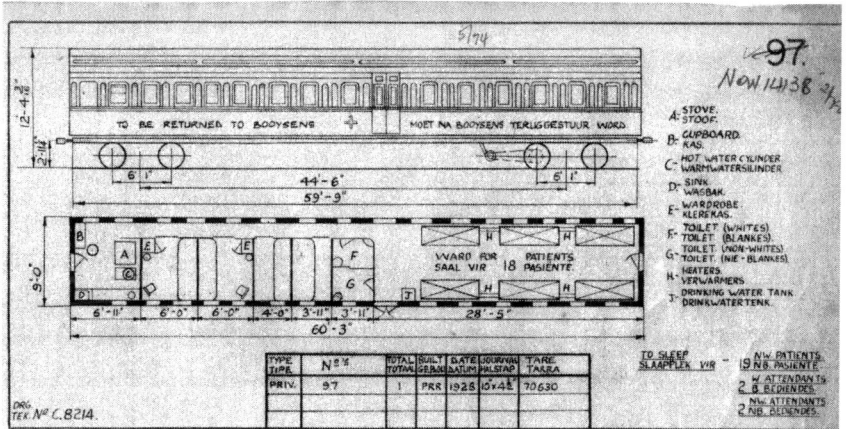

*Technical drawings for SAR hospital coaches.*
*In the top illustration, note the coupe reserved for 'mental cases'.*

As had happened much earlier with the insane, the threat posed by the spread of tuberculosis and other unnamed infectious diseases within the mine compounds among Mozambican miners necessitated special attention from the WNLA shortly before the First World War.[32] From surviving drawings used in the construction of railway coaches, it is clear that, by the late 1920s, the railways were building specially designed 'hospital coaches' for use when evacuating stricken miners from the Rand

and back to the Sul do Save. These coaches had barred windows in order to prevent patients from escaping before they reached their destination at Ressano Garcia. By the 1930s there were several such coaches being used in rotation – each, in turn, being attached to the 307 down that left Booysens station every Thursday afternoon to return the sick and, by definition, still infected and infectious to rural Mozambique.

The three biggest killers – pneumonia, silicosis and tuberculosis – did not, however, exist in perfect isolation, in aetiological silos. Indeed, there were complex cross-cutting influences at work for all black migrants, but more especially so in the case of the deep-level miners from the Sul do Save, that increased the risks of falling prey to one or more of the dread three. Alcoholism, inadequate mine diets, exhaustion and physical stress, for example, not only contributed to high accident rates during the early years of the industry but also reduced migrants' resistance to all manner of other illnesses, including pneumonia, while silicosis was often a precursor to TB. Tuberculosis, in turn, could lead to the infection of other organs.[33]

This set of interlinked occupational diseases, in which Mozambican miners were in the front rank of those who died in an unfolding African disaster, was at its most pronounced from the 1890s until the First World War. Of all the peoples of southern Africa at that time, it was the men, women and children of the Sul do Save who were perhaps most ruthlessly preyed upon by the Four Horsemen of the Apocalypse – Conquest, War, Famine and Death.

One estimate put the number of deaths of all black mineworkers on the Rand between 1902 and 1912, the year in which the *Titanic* sank, at nearly 50 000 – a loss of life equivalent to the sinking of 33 *Titanic*s.[34] At a time of exceedingly limited choice, when much of the westward flow of migrant labour was being driven by *shibalo* pumping, the men of the Sul do Save attempted to come to terms with the devastation through a combination of peasant fatalism and traditional belief. In 1912, some of the one in every four Mozambicans suffering from TB were given to singing a Tsonga song that had a brutally honest conclusion: 'The sun has set for me. Oh, my father, I am dying.' Others, suffering from silicosis and TB, attributed their misfortune to having been snared by the evil spirits of dead miners who lived in worked-out sections of the mine: 'They were captured by these spirits and transmogrified into the zombies called *dlukula* and were forced to labour for several days without pay on a diet of mud.' This portrayal of hell mimicked compound life where, during the early years, black workers might have a day's pay docked and their food rations withheld if they were deemed not to have performed their work underground satisfactorily.[35]

At the close of the 19th century and during the opening decades of the 20th century – and for fully three decades thereafter – the Chamber of Mines did not make any centralised financial allocation for cemeteries. There were no centrally planned burials or systematic marking of graves in laid-out cemeteries for black miners, who were simply interred by their compatriots on the various mine properties where they had expired. Death was integral to the labour migrant system, but the death of black miners was treated largely as collateral damage in an industry that seldom lost its ability to focus on profit to the exclusion of almost any other consideration.

In 1899, the Johannesburg City Council, responding to complaints by white citizens who knew about the body-consuming propensity of an industry that many African workers likened to the practice of 'cannibalism', made its reservations clear to the Chamber of Mines when black corpses, barely covered, were transported along public thoroughfares in open carts. The agreed solution, as in the case of ill and maimed black miners – as we will examine more closely in due course – was to attempt to conceal the system as best possible from the eyes of any squeamish white citizens. The corpses would no longer be conveyed on public roads and 'the natives would be buried on the most isolated portion of the mine property'.[36] Discretion and secrecy – designed and maintained by the rich and powerful so as to help keep everyday racial injustices out of the public eye as far as possible – contributed to the development of a white South African electorate whose ethically impaired and contrived everyday social vision laid some of the political groundwork for the increasingly formal systems of first segregation and much later apartheid.

The mining industry had no use for the dead. The disposal of corpses was largely cost-free for the WNLA and the Chamber of Mines in an industry where death was a central feature. The evidence before, and response to, the Native Grievances Inquiry of 1913–1914 illustrates the attitude of the mining industry and the state to questions of death, burial and cemeteries for black workers whose underlying health, disposable income and wages were all simultaneously at risk and largely declining for at least a decade thereafter:

> A certain number of complaints arise out of matters connected with the burial of dead. The most frequent of these is that no coffins are provided unless paid for by the friends of the deceased. I gather, however, from witnesses acquainted with native customs that [it is only among] a minority of natives of the class from which mine labourers are drawn that coffins are used for the burial of those who die at their kraals. In these circumstances and in the absence of any stipulation on the subject in their contracts, I do not consider that this constitutes a grievance.[37]

It is not clear why, if those requiring coffins for burial formed such a small proportion of those who had died, coffins might not have been provided.

'There were some complaints about the native cemeteries on the mines,' the report continued:

> In one case it was said that the graves were too shallow, but this has now been corrected. In some cases old cemeteries have been covered with mining debris; and even those in use are not always properly fenced off and cared for ... Some boys said that they were not allowed to bury their dead themselves; others that they were compelled to bury strangers.[38]

Nor did the rapidly expanding and always profitable mining industry have much use for the living dead – for those who were terminally ill and suffering from silicosis and/or tuberculosis or other infectious diseases. As detection rates improved between the two world wars, those found to be suffering from these or other diseases considered to be physically disabling were systematically shipped out on the 307 down-train, with possibly predictable consequences for the overall state of rural health in southern Mozambique.[39]

Until well into the 1920s, then, the 307 down-train to Ressano Garcia on Thursday afternoons offered terminally ill migrants the most bitter-sweet passage back home imaginable. On the one hand, it held out the hope that they might yet escape the mines in time to make it home, to die among their families, and be accorded the burial rites that would ensure their spirits their rightful place among the ancestors. On the other, it held out the horror of dying on the slow train down and ending up among the truly doomed – those whose bodies and spirits had become disconnected while in limbo.[40]

# CHAPTER SEVEN

# The Down Passage

## Returning the Living Dead

*Today the city gates are at its centre —*
*for its real gates are the railway stations.*
— LE CORBUSIER

The 14-coach 307 down-train complemented the movements of the biweekly up-train that crept into dusty Johannesburg every Wednesday and Saturday morning throughout the year. It will be recalled that the 307 left Booysens every Monday and Thursday afternoon at around 3.15 pm, and was scheduled to reach Ressano Garcia early the following morning, around 7.00 am, for onward connections to Lourenço Marques and Inhambane, a journey that took on average around 16 hours just to reach the border.[1] Up to around 1910–1912, a combination of a few regular coaches supplemented by boxcars and cattle and coal trucks underscored the tendency to treat the 307 as more of a freight than a passenger train. Thereafter the 307 slowly assumed the features of a passenger train of a special kind – the 'Kaffir Mail' – even though it was still treated as more of a freight train, as the drivers negotiated the formal logistical, and unwritten racial and political, codes of the day that informed the functioning of the Eastern Main Line. But, in times marked by an unexpected increase

SOUTH AFRICA

MOÇAMBIQUE

Nelspruit  Komatipoort  Ressano Garcia

Waterval
Belfast  Boven
Pretoria  Middelburg  Waterval
Witbank  Onder  Barberton
Johannesburg  Lourenço
Germiston  Marques

SWAZILAND

0   25   50   75 miles
0   50   100km

—— Railway

**THE EASTERN
MAIN LINE**

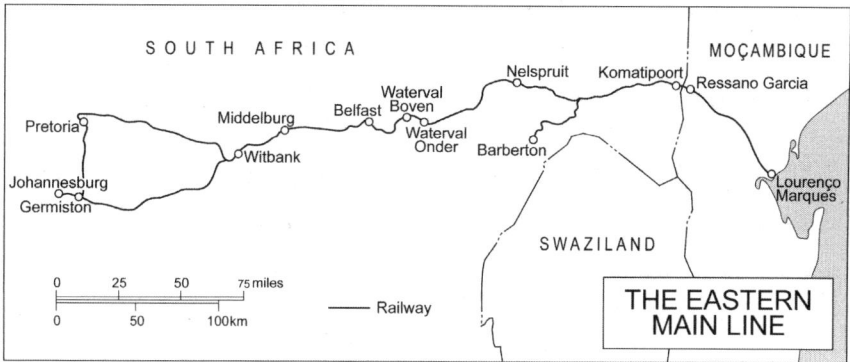

in the flow of migrant labour from, and back to, the Sul do Save – during most of the 1920s – the train regressed as the SAR and WNLA hurriedly supplemented third-class coaches with cattle and/or coal trucks, which, by definition, lacked sufficient supplies of drinking water or necessary ablution facilities.[2]

As one might expect, the operation of the down-train closely resembled that of the up-train insofar as familiar problems such as heating and over-crowding were concerned. But, in truth, there were at least four important differences between the two trains, a few of which might be noted here before proceeding. First, the down-train, the 307, was *descending* from the Highveld plateau to the coastal lowlands and was thus prone to the mirror image of the problem encountered by the ascending 804. Whereas the up-train often slowed to a walking pace as it puffed and pulled its way uphill to the Witwatersrand, the freewheeling downhill 307 could, occasionally, end up speeding as it exited the more congested parts of the system along the far East Rand.

The initial – almost entirely urban – run of the exiting 307 between Booysens and the first major junction, at Germiston, was scheduled to take 30 minutes, but in May 1927, speeding along at 40 miles per hour, the train covered the entire distance in just 16 minutes. This so alarmed the conductor that he was moved to complain, in writing, to his superior at the WNLA – his employer in what was, as we have already noted, a rather unusual private-public partnership with the SAR. 'The coaches were swaying from side to side,' he noted peevishly, 'and I expected to come off the rails every moment.' In prescient fashion, he went on to suggest: 'If this sort of thing continues I predict a bad accident and some score natives being killed to say nothing of your conductor.'[3]

Second – and as we shall explore in more detail further down the path we are following – the outgoing repatriates aboard the 307 differed markedly from the incoming recruits on the 804. It was not just that the up-passengers were probably in better mental and physical shape than were the body-battered and mentally drained down-passengers, but rather that the incoming 804 was, in terms of personal goods and money, dirt-poor while the outgoing 307 was, by comparison, a veritable bank on wheels – something that just about every man, woman and child of every creed and colour along the entire length of the Eastern Main Line was fully aware of.

Third – and as we shall return to later – it was precisely because going up to the Rand was a body-rich, cash-poor endeavour, while the home-ward-bound leg was a mind-and-body-poor but cash-flush experience, that the 307 down had far more, and notably longer, scheduled stops than did the 804 up. Nobody had time for a penniless peasant, and on the way up the station platforms were largely deserted and the train gaily waved through. But everyone loved a cheerful proletarian with a money-belt, and in order to get at the cash, the train had to be brought to a halt and its departure delayed as long as possible so that an apparently welcoming throng of traders could take the opportunity to talk workers into parting with their hard-won earnings.

Fourth, for some time before, and especially after, the First World War

– more or less a decade after the worst of the industrial slaughter on the Rand had passed and the death rate of black miners from occupational diseases was declining comparatively rapidly – the WNLA began to make greater use of the specially fitted 'hospital coaches' for the removal and repatriation of the many living dead and maimed and a few of the manifestly insane back to Mozambique.

While this constituted a clear improvement on the status quo ante for most accident victims, the terminally ill and other patients being transported back to the open-air clinic that parts of the Sul do Save had become, it was not entirely unproblematic. While the Red Cross emblazoned on the sides of the hospital coaches guaranteed patients a measure of protection while the train was in slow motion moving east down the line, they also provided the many platform parasites and a few train robbers with a way of knowing where exactly the most vulnerable of all were housed whenever the 307 was stationary.

Right from the turn of the century, and probably since the mid-1890s, the Chamber of Mines had increasingly been aware of the fact that the rapidly developing deep-level mines and the mine compounds were exacting an extraordinary mental and physical price from African miners, and more especially so from the highly regarded Shangaan-Thonga labour force drawn from southern Mozambique. Within weeks of establishing its new recruiting arm, the WNLA, in 1902, the Chamber placed an order with a United Kingdom supplier for the first of many hospital coaches that would be attached to the regular CSAR train that ferried the physically impaired, those suffering from infectious diseases and the terminally ill to Ressano Garcia. That first coach, No 94, was assembled locally, in 1903, and from a surviving technical drawing it would appear that years later, it was converted and refitted to become a 'workman's coach'.

Such evidence as has survived – again, taken from technical drawings

– suggests that it was only shortly after the First World War, and perhaps even as late as the mid- to late 1920s, that several additional, separate, dedicated hospital coaches, capable of being locked down and fitted with barred windows throughout so as to prevent the escape of any patients, were constructed and used for the repatriation of those suffering from infectious diseases. This meant that the 307 train down to Komatipoort/Ressano Garcia each Thursday afternoon had not one but often two hospital coaches attached to the rear – one for those suffering from infectious diseases, primarily TB, and the other for non-infectious cases who were nevertheless very seriously incapacitated, including one or two who were insane and locked into a coupe that had a window fitted with sturdy metal bars.

Before the war, amputees, accident victims and those suffering from contagious diseases, along with the terminally ill, were separated from the mass of returning migrants on the rest of the train but could also sometimes be transported back to Ressano Garcia in the familiar combination of boxcars and cattle or coal trucks.[4] Among the returning sick, unsentimentally termed 'rejects' by the mines and WNLA in an era when modern manufacturing industries routinely dealt with 'factory faults', the primary distinction was between 'lying down' cases, sometimes suspended in hammocks, or 'stretcher cases' and those patients capable of sitting.[5] A 'lying down' case referred, in the first instance, to the patient's medical condition but, on a train that could often be overcrowded with standing room only, it was also a meaningful concession in other ways.

'Mental cases' consisted of one or two psychotics, potentially violent, and therefore in need of restraint, who were placed in a compartment where the WNLA's European conductors, using 'makeshift keys', made certain they were locked down for the journey to the border. The remaining, more numerous cases manifesting signs of various psychological traumas, which remained otherwise unrecorded anywhere, were grouped together as being 'mentally deficient' only because they had retreated into a world of their own and were markedly less physically active and therefore posed no immediate threat. But they too were often sufficiently dysfunctional to be locked down and kept out of the public eye for the duration of the protracted wait at the busy, unpredictable junction at Germiston.[6]

Seen from the outside by skittish white commuters with little appetite for confronting the workings of the system head-on, the hospital coach with its painted Red Cross offered – as WNLA officials knew only too well – a reassuringly modern image of industrial care and concern for black workers. When considerations of cost finally prompted the withdrawal of the standard hospital coaches, in 1963, there was some concern that the absence of such a facility and the sight of 'sick mine repatriates' in a 'weak and debilitated condition' detraining at Ressano Garcia might lead to 'adverse publicity'.[7] In Mozambique, the local WNLA manager had good reason to be nervous. Ten years earlier there had been a public outcry when disabled and diseased patients were seen being carried through the streets of Lourenço Marques on stretchers before being placed in an open van for transport to the docks for shipment back to Inhambane. The solution suggested at the time was to make compassion manifest by purchasing a second-hand ambulance that would be better suited to moving the unseen maimed occupants.[8] The sad truth was, however, as we shall note in due course, that the term 'hospital coach' was always a bit of a misnomer. The crude facility was more of a rudimentary, self-contained dormitory-on-wheels for the seriously or terminally ill than a hospital.

A few of the earliest hospital coaches, in use by 1922–1923, had all the compartments bar a coupe removed from a third-class saloon, leaving a large rectangular space, 'the ward', bisected by an aisle that allowed for the easy manoeuvring of any stretcher entering via the small balcony at the far end of the coach. Each of the walls thus created housed a tier of three bunks made of woven wire fitted with loose cushions suitable for recumbent cases, including those who had lost limbs or who were suffering from respiratory diseases. The window in the coupe, at the conductor's end of the coach (the rear), as we have noted, was fitted with iron bars and a lock 'on the door which cannot be operated from inside of the coupe' in order to secure potentially violent and inert 'mental cases' alike. Adjacent to the coupe, toilets were installed for the use of the supposedly 'hospitalised' patients.[9]

Two custom-made hospital coaches – one with 12 berths and the other with 18 – were refitted, with minor modifications, every 15 years or so, but the original template remained largely unchanged for the next 40 years. The upgrading envisaged in 1938 made provision, for the first time, for the use of sliding doors to obviate the old practice of passing 'stretcher cases' through windows, but the outbreak of the war meant that the new coaches entered service only in late 1940.[10] The 1951 refit retained all the familiar features but it was stipulated that the sliding doors on each of the coaches be provided with large metal staples so that they might be locked with WNLA-supplied padlocks. The WNLA also wanted more work done on one of the four compartments in each coach that would be used for holding the 'mental cases'. Light switches had to be located outside the doors, internal fittings such as mirrors, racks and washbasins had to be removed, and, as always, heavy iron bars were to be fitted to the windows. By early 1963, the last of the hospital coaches were once again making use of coupes for restraining the mentally ill who were potentially dangerous.[11]

Regardless of how well or poorly these hospital coaches were constructed, their 'wards' equipped, or the patients tended to, in the end all were hooked up to the rear of the 307 and were therefore forced to share in the trials and tribulations that came with the 'normal' operating realities of the down-train. WNLA officials might have hoped that, with the Red Cross emblazoned across the sides of the coaches, the eastbound 307 might lay even greater claim to its status as a black 'passenger' service than its westbound equivalent. But, for many of the poor whites who dominated the lower ranks of the SAR system, including the driving, marshalling and signalling, the train remained, in essence, just another 'Kaffir train'.

The 'coach for sick boys' having 'no drinking water owing to pumps being out of order' may have been a one-off complaint by a frustrated conductor and the men, but other, more serious problems sometimes persisted for months on end. Speeding across the most congested parts of the down-run meant that 'the sick boys had all they could do to keep in bed by holding on with both hands'. Railway operatives, under pressure to keep supplies of bunker coal moving from Witbank to the port at Lourenço Marques, seldom resisted the temptation to add backed-up coal trucks to the 307 when it arrived at Waterval Boven between midnight and 2.00 am. The procedure necessitated a good deal of shunting back and forth in order to move the hospital coaches to the rear of the train so as to accommodate the coal trucks higher up in the sequence: 'Several sick natives were thrown out of their bunks to the floor and severely shaken by reason of the violent jerking'. Adding weight to the train in this way, in midwinter, at one of the coldest points of the journey, meant that the Class 14A locomotives had difficulty in generating sufficient pressure to haul the load, resulting in the steam heating to the passenger coaches being turned off. So severe and persistent were these potentially fatal disruptions that, in 1933, the Chamber of Mines – not much given to compassion – twice complained to the SAR General Manager about the 'gross ill-usage of medically unfit natives and the great discomfort inflicted on other passengers resulting from the attachment of coal traffic' at the midpoint, Waterval Boven.[12]

The addition of coal trucks to the 'Kaffir Mail' while hauling Red Cross coaches turned it into a 'mixed train', removing its 'passenger train' status and thereby rendering it all the more vulnerable to delays occasioned by unsympathetic signal-box operators. European conductors in charge of the 307 and WNLA doctors back in Booysens – spoken of in derisory terms by others in the profession as 'Kaffir doctors' – were aware of the resulting dangers.[13] The so-called hospital coaches were never designed or operated in a way to guarantee that the condition of the patients remained reasonably stable in the day and a half spent on the rails, let alone to facilitate any healing process. Few, if any, patients ever left a Red Cross coach in a steady state, or in a better condition than they entered it. As one of the European conductors twice conceded within the space of a few months in late 1920:

> [It] is *essential* that sick natives who travel by this train should reach Ressano Garcia with as little delay as possible, as *naturally the train journey has an ill effect on these natives*; therefore it is important that the time occupied in reaching their destination should be kept to the minimum.[14]

> Moreover, the sick boys should be got to Ressano Garcia as soon as possible on Friday morning for as a rule some of them are usually in *a pretty bad way by the time we arrive at Kaapmuiden* after a night's shaking up in the train.[15]

By and large, then, the earliest hospital coaches had a noticeably deleterious effect on the health of men who had already been seriously, if not fatally, compromised by mine labour. The men of the Sul do Save on the 307 knew they were involved in a race against the clock, hoping to get home in time to see a traditional healer or, at least, to die among kin so as to be assured of a proper burial, and that by taking part in the race against death, they risked expiring on the train. So, too, the Chief Medical Officer back in Johannesburg knew what the priorities were. 'If we attempted to detain all natives of this class who were not absolutely fit to travel, apart from the fact that we would soon be overcrowded,' he noted in 1923, 'there would be continual dissatisfaction and complaints, not only up here, but in the [labour-supplying] territories.'[16] For the WNLA, the Hippocratic Oath was often left in the balance, with logistics and institutional reputation sometimes outweighing considerations of lives and suffering.

Officials from the state's Department of Native Affairs, too, knew that the shuttle service back to the Mozambican border was, as they put it, little more than a 'gamble with death'. 'In many cases Natives have no hope of recovery and they are most anxious to leave for home on the first possible occasion,' wrote one inspector. 'The Medical Officer patches them up to the best of his ability so they will be able to stand the train journey.'[17] The Eastern Main Line enabled the Chamber of Mines and the state to help keep the Rand clear of corpses and export expiring bodies to the Sul do Save.

In a sense, then, for a decade or so before the development of improved mine hospital facilities on the Rand, Red Cross coaches were little more than coffins-on-wheels and part of a systematic mass-evacuation campaign for permanently disabled 'rejects' and the terminally ill. Nor was the evacuation from Booysens easily effected, because the slope at a station lacking a proper platform did not allow 'stretcher and crutch cases' to get closer than 'about 200 yards from the coach conveying them to Ressano Garcia'. Even then, as noted, for many years some desperately ill or disabled patients had to be lifted up before being passed into coaches through the windows.[18]

Like all returning migrants, patients in the hospital coaches were given two loaves of bread for the long journey home but afforded extra consideration by way of an additional ration, of 'beef and jam' and, where necessary, a tot of brandy to ease their pain.[19] In the early 1920s repatriated colliery workers, handed an additional biscuit for the trip to Ressano Garcia while the down-train paused at Witbank, apparently had every right to be envious of those in the Red Cross section because, it was said, the biscuits were 'unfit for human food'. 'In fact,' reported one conductor, 'I doubt if they are fit for pigs.'[20] An enduring complaint over four decades was that passengers were expected to negotiate poorly lit 'wards' or compartments of regular saloons, which, along with periodic overcrowding, constituted yet another hazard on the 307.[21]

The mass evacuation of the living dead from Booysens to Ressano Garcia, Lourenço Marques and Inhambane lasted from about 1901 until well into the late 1920s when the three biggest killers in the compounds – pneumonia, silicosis and tuberculosis – were at last being brought to heel. By then, however, an unknown number of men, many without their names being easily retrieved let alone systematically recorded, had died aboard, first, the boxcars and cattle and coal trucks and then, later, in the hospital coaches. How many exactly died on the train while the 307 service was at its most basic and killer diseases at their most rampant, between 1902 and around 1914, is unknown. A cursory examination of a set of hopelessly incomplete records covering the period 1918–1927, when the system was improving, suggests that more than a dozen migrants, who can fairly easily be accounted for, expired on the Booysens–Ressano Garcia leg of the journey to the border, and an unknown number between there and Lourenço Marques.[22] Conservatively estimated, the number of deaths on the 307 from 1902 to 1963 might have run into scores, if not hundreds.

Meeting death on the 307 was, for the most part, a brutal and lonely experience. By the early 1920s, when, for reasons that are not clear, such occurrences were briefly better reported on, one or more bodies of black miners were retrieved from the train at every major halt along on the Eastern Main Line after Witbank – at Waterval Boven, Nelspruit, Kaapmuiden, Komatipoort and Ressano Garcia.[23] In 1925, the Compound Manager at Booysens had to be pointedly reminded by the Chamber of Mines that putting the living dead on board the 307 was not an acceptable way of disguising fatalities whose origins clearly lay on the Witwatersrand: 'As regards natives who pass through your hands for repatriation and die on the trains, such deaths will, as heretofore, be regarded as compound deaths, and are to be recorded in your figures accordingly.'[24]

It would appear, however, that many of those who survived the down journey on the 307 were in such appalling condition by the time the train pulled into the station at Ressano Garcia or, less frequently, Lourenço Marques that they were simply moved into the WNLA hospital, where they expired. Again, it is difficult to know the exact numbers – the records are incomplete – but some idea of the scale can be gleaned from the 'Death Returns' that were submitted for a few months in the mid-1920s. At that time there were seldom fewer than two deaths a month among repatriates who detrained at the border post. In May (winter) 1925, there were nine such deaths at Ressano Garcia; in July, two; in September, two; and in October, four. In February 1927 (late summer), the Chamber of Mines informed the Portuguese Curator of Natives in Johannesburg that 'the first five natives [of six] detailed in your letter died after arrival at Ressano Garcia'.[25] By the early 1930s the position appears to have improved, with far fewer fatalities being recorded in the morbidly designated 'Death Returns'.[26]

There was not much room for compassion or sentimentality when it came to train deaths. 'I transhipped some of the sick boys who were very overcrowded on leaving Booysens,' one of the more humane WNLA conductors dutifully recorded in 1922, and then immediately moved on to conclude his report in the customary businesslike fashion: 'Reject Disc Number 5852 died en-route and the police at Komatipoort took charge of the body.'[27] From the same official, in January 1925: 'I beg to report that Reject Disc No 8339 died en-route to Ressano Garcia on the morning of the 13th instant. Body removed from train at Komatipoort and 75 pounds returned herewith.'[28] In most cases, however, state officials merely noted that an individual bearing an incomplete name of sorts had been 'found dead' on the train. It was a formulation that suggested that, in his final moments, the dying man had been on his own only to be discovered dead sometime later.[29]

District surgeons conducted post mortem examinations of bodies that had been passed on to the police. 'Native Sombane, Portuguese Passport Number 471954, Z 2186,' boarded the 307 at Modrea, on the East Rand, in 1923, 'vomiting blood' and was 'found dead' shortly after the train left Kaapmuiden. 'As a result of a P/M examination the cause of death was given as Silicosis, Tuberculosis and Pneumonia,' but, noted an official reporting on the case, 'I regret I am not in a position to state whether the lungs were sent to the Miners' Phthisis Bureau' for purposes of medical research.[30] Unclaimed bodies, interred in paupers' graves somewhere along the Eastern Main Line, were clearly not always necessarily intact.

The experiences of Sul do Save men aboard the 307 and 804 were part and parcel of an extended circuit embodying crass processes of dehumanisation that, for those among a minority of fully healthy returnees, usually started and ended at Ressano Garcia. And, just as those aboard the up-train were somehow never considered fit to be called *miners*, so those on the return journey travelling in hospital coaches were, like so many indisposed small children, termed 'sick boys', or 'sick natives' but never qualified as *patients*. How could they be? For two decades before the Red Cross coaches came into service in the early 1920s until four decades thereafter – that is, for more than half of the 20th century – not once were the hospital coaches graced by the presence of so much as one of the dozen WNLA doctors who had willingly declared returning migrants 'fit to travel', and that included those cases where the patients had died.[31]

These, then, would have been some of the very same doctors who, over the years, were party to attempts 'to convert pulmonary tuberculosis in miners from an occupational disease into a community acquired, non-occupational disease', with the attendant 'shedding of responsibility that would go with such a transformation'.[32] For at least some WNLA doctors and the Chamber of Mines, certain important occupational diseases emanated in the countryside and community rather than in the mine compounds or collieries, and the diseased black miners could therefore be returned to sender with impunity.

The standard of 'fitness to travel', complained the WNLA's District Manager in Lourenço Marques, in a 'confidential' letter in 1919, is 'very low'. Many of the repatriates who made it as far as Ressano Garcia 'are unable to travel further', while in other cases returning Mozambican miners 'were carried to their homes in hammocks'. And, he continued, rather tentatively, knowing that he had to tread carefully because he was not 'a duly qualified medical man' and was offering a lay opinion in a highly bureaucratised system, 'probably very few such cases ever recover'.[33] The 307 was, in part, the most visible moving component of an interlocking system for the export of death.

During the six-decade-long absence of a qualified medical man on the down-train carrying between 400 and 700 returning migrants, along with a Thursday quota that might include amputees, the blind, cripples, the clinically depressed, epileptics or the occasional violent psychotic, as well as those suffering from pneumonia, silicosis and tuberculosis, the responsibility for the passenger operation of the 307 at any one time rested almost entirely on the shoulders of a single European WNLA conductor. Before the mid-1930s, none of those in the long-serving but small complement of regular conductors had any formal medical training or qualifications, but when they retired they were replaced by one or two men who appeared to have been at least partially trained as nursing orderlies.[34]

The Chamber of Mines and the readers of the English-medium press on the Witwatersrand were content to see the hard-pressed, unqualified and occasionally well-meaning conductors portrayed as 'hospital orderlies' – in line with the display of the Red Cross on coaches. They were, for the most part, totally out of their depth when it came to dealing with terminally ill or seriously sick patients in hospital coaches. A series of rail accidents towards the end of the First World War allowed two of the longest-serving conductors to point out 'the futility of coping with an accident of any size with the present First Aid means at our disposal', and to plead for additional medical supplies including a 'stove, so that [an] ample supply of hot water might be available for washing wounds, etc.' And, in case their WNLA managers failed to understand the fragility of the entire system within which they were expected to operate, the conductors reminded them that there was not a single qualified doctor to be found on the entire hazardous 206-kilometre stretch between Waterval Boven and Komatipoort, and that in the event of any accident there, the nearest professional help was at least six hours away.[35]

Fifteen years later, the level of service provided to patients in the Red Cross coaches was still rudimentary in the extreme, given the range and nature of ailments that conductors were called upon to deal with during the long night runs to the border. In a letter to the General Manager of the SAR, however, the Secretary to the Chamber of Mines unashamedly buffed the image of what he cast as the entirely commendable, powerfully humanitarian endeavour on the part of its WNLA subsidiary, which contrasted sharply with the consideration given to the 307 by a grubby, profit-seeking, state-owned enterprise. Some of the Johannesburg industrialists, men of a supposedly 'progressive' stripe, better-educated English-speaking urban sophisticates with supposedly more enlightened 'liberal' views, were content to chide or dismiss Afrikaans-speaking railway administrators and poor white workers as 'backvelders' and 'wheel-tappers', while out at Ressano Garcia their own employees, the black living dead extruded from coal and gold mines, were being hauled through the coastal bush in hammocks and readied for the loneliest journey of all. Exposed to the transformative agency of time, the arrogance, condescending nature and deceits of class and culture – first etched in the palest tinctures of business logic and pure reasonableness – assume the dark hue that characterises obscene graffiti left staining the walls of abandoned alleyways in what once was a thriving colonial economy:

> Whilst the Association [WNLA] does all that is humanly possible to render the conveyance of native convalescents as comfortable as can be by way of providing a European conductor, qualified in first-aid, bedding and medical comforts, and generally arranges the whole of the profitable native traffic carried by these trains with a minimum of trouble to your officials, the Railways Administration for its part is apparently not prepared to grant a reasonable *quid pro quo*, and is inclined to give less consideration to the running management of these trains [on the Eastern Main Line] than is given to a perishable freight train.[36]

The magic of managerial words, deftly deployed, turned the ill, the walking wounded, the lunatics and others into 'convalescents' – who by the WNLA's own admission left the hospital coaches in worse condition than they had entered them. It was precisely part of this same unspoken, 40-year-long tussle between the cash-starved, revenue-seeking Railway Administration on the one hand, and the profit-driven, cost-paring initiatives of the Chamber of Mines on the other, that eventually led to the partial abandonment of the WNLA's already minimal so-called Red Cross service.

Within months of the introduction of new custom-built hospital coaches in 1923, the carriages – or, more accurately, their costing – risked becoming a bone of contention between the management of the Chamber of Mines and that of the SAR. The two administrations were linked in a private-public partnership ostensibly in the 'national interest', but neither party wanted to be left out of pocket for the hospital coaches. The problem was that while the WNLA paid the railways a per capita fee for every recruit or repatriate carried on the often jam-packed third-class saloons of the 307 and 804, the Red Cross coaches were, by comparison, sparsely occupied and therefore, seen in strictly accounting terms from the perspective of the SAR, largely unprofitable. The Chamber of Mines and the WNLA, on the other hand, needed the hospital coaches to be part of the public face of modern industrial welfare but took the view that any deficit faced by the railways as a result of the Red Cross coaches was more than compensated for by the huge stream of revenue generated from the regular passenger coaches on the up- and down-trains. It was, in short, a dispute about accounting between an arm of the state committed to maximising revenue and the biggest business in the country, which for many decades was not particularly deeply invested in running a service for 'convalescents', many of whom, in truth, would never convalesce.

At a time when an ascendant mining industry was central to the economic well-being of the country, the SAR was never really in a position to impose its accounting will on the Chamber of Mines with regard to the hospital coaches and, in effect, subsidised the service by absorbing the loss in revenue that came from evacuating those southern Mozambican casualties broken in body and mind, or who were terminally ill. But businesses come and go, while the state is enduring. By the early 1960s, with the Afrikaner nationalist government having triumphed in its long struggle for a republic, and intent on implementing its grand apartheid policies ever more firmly, the moment arrived to test just how committed the conflict-avoiding, cost-paring Chamber of Mines was to running a quasi-medical service for 'foreign natives' drawn from a neighbouring state.

In late 1962, the SAR gave notice that, no longer content with revenue obtained solely from per capita passenger fees, it intended to impose a surcharge on the hospital coaches – something that came as 'a surprise' to the WNLA. The appeal that resulted from this sudden imposition of new fees was brushed aside by the SAR in the no-nonsense approach much favoured by a tough-minded government with fascist tendencies. 'It is not denied that considerable revenue is derived from the conveyance of Bantu mine workers but this in itself does not warrant their obtaining preferential treatment,' wrote the Acting General Manager of the SAR, 'and the mere fact that these coaches have up to now been included in the trains mentioned above does not justify the continuance of this service free of charge.' Black men being shipped out of the industry in parlous mental or physical condition could not, under any circumstances, be converted into what, in strictly accounting terms, might be seen as 'preferential treatment'. The WNLA was then faced with an awkward choice: it would either have to pay the full surcharge on the coaches to the SAR in order to benefit from what, in large part, had always been a public relations exercise bearing a Red Cross, or perhaps see the coaches withdrawn entirely.

The WNLA manager in Lourenço Marques, where the news of the withdrawal of the Red Cross service might be most negatively received among the men of the Sul do Save, along with a few fidgety Portuguese residents in the city, had to be suitably primed for the masterly managerial compromise that was being driven from the Witwatersrand. 'You will, I am sure, be very sorry,' the General Manager wrote to his coastal colleague, 'to learn that the hospital coach which has been attached to the labour train running between Johannesburg and Ressano Garcia will be taken off, except in those cases where it is essential that natives be sent home by the hospital coach.'[37] The odds in the 'gamble with death' had just shortened dramatically; henceforth only the very worst cases would have the luxury of a supposed Red Cross evacuation. And, in keeping with the spirit of the new arrangement, the two remaining white conductors were redeployed to the WNLA hospital in Johannesburg and their places taken by two new, full-time 'non-European' conductors.[38] For big business in South Africa, as in most parts of the world, politics and economics almost always reached an agreeable, very practical compromise.

By mid-1963 the full down-train, replete with two Red Cross coaches, a ghostly spectre most often glimpsed gnawing away at the tracks of the great Eastern Main Line during the early hours of the morning, was seen less often. The 307 and its hospital-on-wheels without doctors or patients continued to do as it had done for half a century, ferrying the living dead and the mentally impaired across the River Styx at Komatipoort/Ressano Garcia. The down-train had often left the souls uncoupled from the bodies of the returning migrants, but that was not its full repertoire. It was also the vehicle by which men were separated from their savings and few worldly goods.

CHAPTER EIGHT

# Running the South African Gauntlet

## Asset-Stripping Returning Mozambican Migrants

*Through the dark night chasing the morning light,*
*that headlight streaming white through the night.*
— RICHARD RATCLIFF

The ostensible purpose of the 307 down-train was to convey contract-expired workers, mine 'rejects' and other 'sick boys' along with all their personal possessions and cash savings back home to the Sul do Save as expeditiously, safely and securely as possible. But, as men from the Sul do Save well knew, the journey home was unlikely ever to be comfortable. Repatriates, sensing the physical challenges to be expected when travelling on WNLA trains, had even more reason than usual for wanting to get home to their families and villages. Insofar as it related to the few who were clearly insane or the many more bunk-bound terminally ill evacuees housed in the hospital coaches, the need for a swift passage back east was an expectation shared by conductors who knew that unnecessary delays only increased the chances of having to 'gamble with death'. There was then a *hope*, reasonably widely shared up and down the corridors of the train, that the 307 would do precisely what steam locomotives were designed to do – conquer distance through power and *speed*.

As ever, however, the icy water of economic reality provided a rude awakening for anyone lost in a dream world where hope might be coupled to speed. The inward-bound 804, after dumping its quota of men for the collieries in Witbank, deposited all remaining WNLA recruits in centralised Booysens for bureaucratic processing and medical examination prior to distributing them to the mines up and down the Reef. Recalling all contract-expired workers to Booysens before getting them to board the 307 down-train would, however, have been a waste of time and money since the train could easily make various stops as it exited the Witwatersrand and headed out along the Eastern Main Line. Repatriates from gold mines to the west of Johannesburg might gather at Booysens and board the train there, but for those drawn from mines on the near East Rand, or from the Eastern Transvaal collieries, the train halted at Germiston, Modrea and Witbank.

The 307 eased out of Booysens at 3.15 pm on Mondays and Thursdays but, because of the additional stops at Germiston, the largest rail junction in the country, and Modrea eventually reached Witbank only in late afternoon. There, the repatriates detrained for a half-hour march to the WNLA collieries' compound, where they were provided with their last 'free' ration of food and drink for the onward journey. A 30-minute march back to the train and reboarding chewed up yet more time before the 307 eventually slithered out into the darkness for the long haul across the Highveld, reaching the mountains at Waterval Boven between midnight and 2.00 am. There was another lengthy wait here while locomotives were changed for the steep drop into the Devil's Throat before the long, slow descent into the Lowveld where, at various times during the years covered here, there were sometimes scheduled halts at Nelspruit and/or Kaapmuiden, occasioning yet more waiting. From here, it could be relatively plain sailing – if all went well. The 307 crawled into Komatipoort at sunrise, around 7.00 am, having taken 16 to 18 hours to cover the 598 kilometres separating Booysens from the border complex at an average speed of little more than 35 kilometres an hour. Somewhere out there, on a winter's night on the 307 chugging its way slowly east, hope had taken leave of speed.

Set against the inherent promise of the 19th-century steam locomotive, then, both the up- and down-trains in the WNLA system, striving to deliver and return cheap black labour with clock-bound, industrial efficiency to the mines, were severely compromised whenever it came to speed conquering distance. It was the speed of the train, constantly inhibited by the signal-box operators, that lay at the heart of the never-ending WNLA complaints about the 804 and 307 being treated as freight rather than passenger trains. While much of the slowness in the system could be attributed to technical constraints caused by the steep change in gradient and consequent slowness through the mountains, four, sometimes six, lengthy stops contributed significantly to stretching the journey. The system, and more especially the return leg back to the Komatipoort/ Ressano Garcia border complex, was never a smooth, seamless circuit – as implied in the use of the word 'line'. For the 307 down-train, which exited to the east via the Witwatersrand gold mines before joining the main line proper at Witbank, the Eastern Main Line was more deformed insect than snail, a mutant thing with three stubby little legs, with the stations acting as articulating joints.

The stations – and especially the tortuous, twinned border crossings at Komatipoort-Ressano Garcia – acted as severe disrupters, as joints in the legs, breaking the integrity of the journey, bringing the train to a halt, disturbing the composure of the expectant passengers, inducing a feeling of vulnerability in the minds of the peasant-proletarian migrants. Because the station stops destroyed the otherwise suspended rhythms of time and space while the train was in motion, they rendered the miners, their money and their goods temporarily vulnerable to predation. As the peerless Lewis Hyde suggests, 'eternals are vulnerable at their joints', and to get at a god (as in, say, Achilles' heel) or an ideal, he argues, '*go for the joints*'.[1] And go for it they did. As we shall note below, stations up and down the line, from Booysens to Komatipoort and beyond, were the favoured hunting ground for confidence tricksters, fraudsters and petty thieves tracking the homeward-bound 307.

A lack of formal education and familiarity with some of the everyday predations and vagaries of African township life, supplemented by contract-long seclusion in mine compounds, underscored some of the 'naivety' and vulnerability of Mozambican repatriates expected to circumvent the treacherous whirlpools of humanity to be found on the platforms of large stations for the first time. But, in order to understand this purported naivety and vulnerability fully, and in relation to the operation of one type of confidence trickster in particular, it is necessary to understand first how the quasi-military template of uniformed authority imposed on the 'East Coast Boys' and other miners helped soften them up for predatory deception.

From the instant that they first encountered WNLA recruiters and African policemen in the deep rural areas of the Sul do Save until the moment they completed the border formalities upon their return to Mozambique at Ressano Garcia, 'East Coast Boys' were – at various points or times and with varying degrees of intensity and urgency – assaulted, corralled, disciplined, deprived, documented, examined, fed, fingerprinted, guided, incarcerated, marched, measured, marshalled, questioned, searched, raided and released by black and white men in uniform. The state's uniformed men, whose power appeared to be arbitrary and beyond any question, controlled and patrolled the external perimeter, the boundaries of the system, which to the innocent appeared to be a 'normal' part of any citizen's domain. But, in truth, the customs and immigration officers, stationmasters, railway officials and even the Portuguese Curator of Natives on the Rand were, in reality, men whose function, in part, was to assist in concealing the system as best possible from colonial society at large. And, deep within the greater system, the inner workings of the repressive labour regime were enforced by yet more uniformed men. Within the compounds, on mine property and, by the 1950s, even deep down the shaft itself, uniforms were omnipresent. There were 'compound police', 'mine police' and eventually even 'underground police'. Every visual and subliminal code that Mozambicans ever encountered via uniformed men demanded compliance, deference, obedience and silent submission. It was this extended mental panel-beating undergone over 18 months of contract labour and enforceable by criminal penalties that helped render them 'gullible', 'obedient' or 'naive' when faced with confidence tricksters in uniform – black men, white men, and sometimes black and white men in tandem – on railway station platforms.

East Rand stations in particular constituted zones of overlapping, if not confusing, authority since it was at those congested junctions that the normally well-secluded migrants from the Sul do Save could not be fully, or easily, separated from those members of the South African public bent on predatory activities of various types. Mine police escorted repatriates onto the platforms, but at that point their authority effectively lapsed, since the stations themselves formed part of railway property where the enforcement of public order and safety fell to the SARP, which was answerable only to the SAR administration. It was at this point of transmission – at this joint in the system – that the limits of the private-public, WNLA-SAR partnership were yet again exposed, leaving the migrants vulnerable. Who precisely was responsible for protecting the migrants and their wages and possessions on the platforms? Was it the mine police or the SARP, or both? In which case, as so often happens in unresolved territorial disputes about spheres of authority, could it be that the problem often fell between the cracks?

The jurisdictional crack did not end at the platform edge; it extended down the entire length of the Eastern Mine Line, all the way to Komatipoort, by which point it had long since assumed the proportions of a minor crevasse. Because the 307 down-train was part of a private service, albeit one running on publicly owned tracks, responsibility for the security of the passengers and their possessions lay, in the first instance, with the WNLA rather than the SAR administration or, more importantly, the often understaffed SARP. This meant that the Sul do Save migrants, although subject to criminal penalties enforced by the South African authorities if found guilty of breaking their labour contacts with the WNLA, were left largely without protection by the very same state if they became the victims of criminal acts while travelling within the country's borders. Everyone was equal before the law and entitled to protection by the state, it would seem, except those already-fingerprinted migrant labourers who happened to be drawn from beyond the national boundaries. And, what the SARP could not, or would not, provide by way of safeguarding the returning miners and their possessions, the Chamber of Mines and the WNLA – two bodies controlling private police of most descriptions when it came to protecting mine property or on-site black labour – were unwilling to supplement. So, just as the 307 hospital coaches were without doctors or trained nurses for the terminally ill, so were vulnerable passengers carrying meaningful amounts of goods and wages on the slow down-train left without the benefit of protection of the South African Railways or any dedicated WNLA train police.

The storage, handling, integrity and retrieval of miners' baggage destined for Ressano Garcia presumably was a long-standing problem, but no records have survived to indicate what the position might have been before 1914. As the number of Mozambican miners on the Rand doubled, however – from 100 000 to 200 000 during the 1920s – the volume of goods being sent to the border increased proportionately, creating logistical problems at the stations, on the 307 and at the border. When there was insufficient space in the luggage van, migrants' boxes were simply 'loaded onto the platforms of coaches' the previous night, leading to thefts and leaving the repatriates 'greatly concerned'.[2] The Portuguese border authorities, as adept at extracting value from anything inanimate belonging to repatriates as the Chamber of Mines was at squeezing physical effort out of the workers' bodies, were quick to spot what they considered to be a promising new development and imposed a duty on what they opportunistically deemed to be 'excess luggage'.[3]

The handling and securing of increasing volumes of luggage in the van at the rear of the train or in the compartments of the third-class saloons slowed down the process of entraining, occasioning further delays to the already speed-compromised train.[4] In order to increase the efficiency of the boarding process, baggage was at first stacked on the platform at Booysens and other stations east the night before the train left, and later stored in what appear to have been rudimentary, temporary left-luggage facilities.[5]

In an attempt to further promote efficiency within the system, the SAR General Manager, overstepping the boundary of the private-public partnership, suggested to the Chamber of Mines that it might consider affixing printed luggage labels to the boxes, bundles and metal trunks most favoured by the repatriates. The response from his correspondent at the Chamber was brutal and swift. 'The journey to Ressano Garcia involves no change of trains,' the Secretary responded, 'and therefore we see no justifiable reason for going to the expense of issuing printed luggage labels, of which many thousands would be required in the course of a year, or for undertaking the amount of work involved in their issue to natives making that journey.'[6] When it came to containing or reducing costs in the mining industry, black labour was always an easy and vulnerable priority.

Lower down the pecking order, SAR employees – African workers – had their own ways of trying to speed up the departure of the sluggish 307 carrying little-loved 'foreign natives'. The Mozambicans' baggage was treated with scant respect and hurriedly tossed into the van; once, when the door was left unsecured, the luggage 'spilled along the line'.[7] Contempt for the returning miners' possessions could test a WNLA conductor's diplomatic skills once the train reached the border: 'A number of boxes were broken by the rough treatment they received,' one reported, with the result that 'the natives made quite a demonstration on Tuesday last.'[8] Repatriates, keen to protect their most valuable items, then manoeuvred their baggage into their small compartments aboard the 307. But, unlike the standard second-class – European – saloons, third-class coaches lacked luggage racks, so the migrants instead took to stowing their belongings on the top two bunks in the compartment. In a compartment designed to accommodate six passengers but in practice often housing at least seven workers, this exacerbated the problem of overcrowding by reducing the space available for reclining and forced one or more men to sleep out in the coach corridor.[9]

A quarter of a century later there could still be substantial problems when it came to baggage handling on the down-train. In 1956, a conductor, irritated by yet another frustrating, delayed train departure from Johannesburg, reported eventually leaving 'with 740 natives including 71 rejects' and picking up '45 repats and 8 rejects at Witbank'. But, from the very outset, the down-train was more than three hours late. 'All this delay is on account of the booking of luggage at Booysens station,' he complained.

'It is graded a first-class station but has no proper facilities to deal with the booking of the luggage of say 700 natives in a reasonable time.'[10]

But, back in the 1920s, when the WNLA was pushing the maximum 'output' of workers up and down the Eastern Main Line each week, it was never just a question of whether the luggage of returning migrants bound for the border arrived there intact, but whether it arrived at all. Some of it was 'miscarried' to other, unknown destinations, but more serious, by far, was an epidemic of theft that peaked in 1925. Given the paltry wages earned by most Mozambican miners, the items removed from individual workers' baggage consisted, in the main, simply of inexpensive clothing, but well into the 1930s included a fair number of utterly basic loincloths. And, because these goods were of comparatively low value when measured against prevailing European standards, the losses – each representing months of toil underground – were often ideologically downgraded by officials from theft to 'pilfering' arising from 'tampering'. Responsibility for such 'pilfering' and 'tampering', it was suggested, could in large measure be attributed to the migrants themselves, for not taking sufficient trouble to secure their parcels with strong rope or twine, or to acquire proper, non-standard locks.[11]

The authorities appear to have enjoyed little success in stemming the wave of larceny. Going by the contents of one very bulky WNLA file, there seems to have been only one successful prosecution of an SAR employee: in 1925, a guard was convicted and sentenced to four months' imprisonment with hard labour.[12] So extensive did these thefts become that even the normally complacent Portuguese Curator of Natives in Johannesburg felt pushed to raise the issue of losses sustained by returning migrants at Ressano Garcia. The reply from the Chamber of Mines to the Curator – the latter a title more suited to the Catholic Church, from which it apparently derived – echoed familiar jurisdictional problems about where precisely the boundaries of the private-public partnership lay when it came to the migrants. 'Please advise,' wrote the Secretary of the Chamber of Mines, 'whether it is possible to produce evidence in substantiation of the statement that the pilferage complained of actually occurred while the luggage was in the custody of the Railway Administration, and whether the luggage was booked by the natives concerned before boarding the train.'[13] It was a fool's errand. In the late 1950s there was another spate of thefts, and even in those cases where the Railways Administration did accept responsibility for the loss of goods, it generally refused to pay compensation on items of clothing unless documentary evidence as to the precise value of the goods lost could be provided by workers who were, almost to the man, perfectly illiterate.[14]

The simmering tensions between the SAR and WNLA managements went well beyond the question of the safe movement of baggage to the border. Each time the 307 chugged to a stop at a station, the repatriates were exposed, either through the open windows of the train or, if they disembarked, on the platforms themselves, to all manner of unknown men and women. Some of the people standing about, such as officials and men in uniform, were clearly authorised to be there. Others, members of the public, in ordinary clothes, were of uncertain status. They might be informal – unauthorised – hawkers or, perhaps, just other passengers waiting for trains. But, whatever their status, almost all of those waiting on the platform included some who, like crocodiles gathered at a watering hole waiting for their next meal, knew that the returning miners were carrying cash. For the duration of the train stop, the platforms were converted into open, visible and officially sanctioned or partly concealed illegal trading zones – small but busy marketplaces where opportunity and risk rubbed up against one another and exuded an odour of excitement and uncertainty.

The difficulties faced by the down-train passengers and the WNLA were, however, slightly more complicated than appeared at first glance. The SAR Catering Division, no less so than any of the independent hawkers, was keen to sell as many items of food and drink as possible – if not various small trinkets and toilet necessities – to the repatriates. In theory, the Catering Division enjoyed a monopoly of the local market since the stations formed part of state property. To that end, it could legitimately call on the railway police to keep the platforms free of any meddlesome independent outside hawkers. The SAR therefore always had good reason to try and get the 307 to stop at various junctions, or even sidings, so as to increase its revenue stream. The WNLA management, in turn, wanted to minimise such delays so as to ensure maximum rates of efficiency as it sought to turn around its trains and supplies of cheap black labour.[15] The private part of the private-public partnership wanted time saved; the public part, time spent.

The resulting tension between the partners proved impossible to resolve, precisely because the difficulties once again lay in the joints of the system and were exacerbated by the fact that the railway police were also too thinly stretched to be able to enforce the SAR monopoly fully at all stations, and more especially so at the more numerous and busier stops along the East Rand. Independent hawkers flushed by the SARP from a large junction such as Germiston simply moved one stop up or down the line until it was safe to return to their favoured site, in an endless game of musical chairs. In practice, then, the larger stations in particular often saw a mixture of SAR catering staff and independent hawkers vying for the attention of the 307 passengers and manifesting differing degrees of honesty and urgency as they sought to bully or cajole the men from the Sul do Save into buying their wares. In order to minimise the resulting abuses and exploitation, the WNLA managers had to appeal to the

General Manager of the SAR either to curb the activities of its own cater-ing staff – a battle usually lost, for obvious reasons – or to get it to use the SARP to try and rid the station of independent hawkers, which produced only temporary gains.

Beneath the commercial mist that enveloped the station markets there was, however, a second logic at work that was perhaps most easily detect-able from after the First World War and until the mid-1930s. When the miners, accustomed to an inadequate and insufficiently nutritious diet on the mines, boarded the 307 they were, as ever, hungry but, unusually, also relatively cash-flush. The WNLA provided them with two loaves of bread on their departure, and at Witbank, as we have seen, they were marched to the last supper – a final meal meant to sustain them until lunchtime the following day at far-off Ressano Garcia.

But, as most of the returning migrants already knew from their own experiences on the journey up, or through talk in the mine compounds, what the WNLA gave them was seldom enough to sustain them fully, and dry bread, eaten without jam or relish, held little appeal.[16] They were therefore both able and predisposed to purchase not only one or two last-minute 'luxuries' from the commercial hawks circling the platforms, but also the few supplementary items of food and drink necessary to see them through an uncomfortable journey lasting 16–17 hours. The last-minute acquisition of these items did not, however, necessarily occur in totally random order all the way down the Eastern Main Line. In general, there was a greater propensity on the part of the repatriates to purchase trinkets and toiletries on the East Rand stations before they reached Witbank, where a meal awaited them, and to buy additional food and drink further down the line, as hunger pangs increased and the world slumped into the Lowveld between Waterval Boven and Komatipoort.

This pattern of consumption arose, in part, because hawkers in the urbanised Witwatersrand had easier access to wholesalers and benefited from more competitive prices than did those who had to operate in the smaller centres further down the Eastern Main Line. This meant that, generally speaking, the street-smart hawkers along the East Rand could offer goods at prices more competitive than those further down the line. But there was a flip side to this, because once the same sharp operators sensed that they were dealing with foreign yokels – so many peasant-proletarians who were on their way out of the country and uncertain as to the value of money that could change at the border – they indulged in shameless price-gouging at the expense of the returning Mozambicans.

At Geduld, but more especially Modrea station on the East Rand, problems were experienced with a dozen (a number that eventually grew to around two dozen) male and female Assyrian immigrants that lasted fully half a decade, from 1922 to 1927. The hawkers would arrive at Modrea – a private siding paid for by the mining companies but run by the railways – from Benoni in a horse-and-cart convoy, six deep, and begin 'pestering the natives', leaving repatriates 'grossly overcharged' and frequently without 'their proper change'. Aggressive and reputedly 'pretty free with the knife', the Assyrians appear to have specialised in waiting until the 307 was just about to depart before hurriedly palming off razors costing a shilling each on the Mozambicans, charging them two shillings and sixpence per item, and hoping that, in the ensuing rush to depart, the migrants would forfeit any change due to them. Once only did railway detectives – never quite certain as to which magisterial district or under what jurisdiction the musical-chairs station fell – manage to get a few of the hawkers prosecuted and fined in the Benoni court, but even that failed to act as a deterrent.[17] The East Rand stations were akin to tollgates.

At the huge Germiston station, where there were several African hawkers selling cigarettes and matches and a fair number of 'white lads with trays of stuff going along the train', the railway police were always much keener on enforcing the SAR catering monopoly. But sometimes the SAR staff there, including the black employees, were at least as willing to exploit their 'brothers' as were others yet further down the line. In 1921, when matches were selling at a government-controlled price of eightpence per box, one of the licensed black hawkers was selling them at a shilling a box and pocketing the difference. When he sold a migrant four boxes of matches for four shillings, the equivalent of a day and a half's underground labour, an irate WNLA conductor was moved to appeal to the SAR for the 'more efficient supervision of these native salesmen'. At Witbank, the WNLA had to fight off an attempt by the SAR, in 1936, to institute a further delay of '40 minutes, or any other length of time for the convenience of the SAR Catering Department'.[18]

But it was at Waterval Boven, where during the wee hours of the morning the locomotives were changed and coal trucks added to the 307 for the hazardous descent into the Lowveld, that the sleepy repatriates often had to endure the most unpleasant assault of all on their ears and pockets. Difficulties there, dating back to the time of the First World War, were attributed at various times to a local Asian storekeeper and his black employees, a few independent African hawkers, and half-dozen or so SAR coach cleaners who had the advantage of possessing keys to the carriages. Some idea of those wartime train invasions of the 307 can be gleaned from a letter written by a conductor to his WNLA superior, in 1921, when there were renewed requests to sanction hawking there after the station had been clear for some considerable time:

> If you will look up my old reports on this subject, when the selling of refreshments was in vogue at Waterval Boven, you will see that the method of rousing the boys up was by going along the train and

hammering on the side of the coaches with a stick, and shouting. Moreover the boys were continually complaining of being robbed, and given wrong change. This can easily happen, as the coaches have no lights, and what business is done is practically in darkness.[19]

At the same mountain halt, local black entrepreneurs soon cottoned on to the fact that the down-train carried any number of sickly Mozambicans in various stages of physical and mental disrepair. For at least a decade, and probably much longer, the station was a major site for the selling of traditional African herbal medicines to the walking wounded of the Sul do Save, while simultaneously achieving a certain notoriety as being a place where the hawkers were happy enough to steal from migrants. At Modrea, you could buy city goods such as mirrors and razors, and at Waterval Boven you could purchase country cures, but at both stations you needed to have your wits about you if tendering an amount that required any change.[20]

Below Waterval Boven, as the journey dragged on through the Lowveld with the 307 inching slowly east towards the approaching dawn, many of the repatriates – their last substantial meal by then close on ten hours behind them – experienced renewed pangs of hunger. By then, even the WNLA-issue dry bread was becoming a possibility. First at Nelspruit, and then at Kaapmuiden, these realities were not lost on local traders, or the ubiquitous SAR Catering Division. The Railways Administration canvassed the WNLA on the idea of extending the stopover at Nelspruit, 'to allow the lessee of the refreshment store at that station sufficient time to cater for the natives'.[21] But sometimes the SAR simply rejigged the timetable to its own advantage without first informing the WNLA officially, and it was left to the 307 conductor to spell out what lay behind some new and unexpected delay:

Ostensibly the delay of half-an-hour or more at Kaapmuiden is to allow the Station Staff time to collect tickets, but it appears more than likely that the *real reason* of the delay is to give the Catering Department time to sell a few tins of jam etc. to the boys as I am aware that the man in charge of this Department has for some time been agitating for our train to be held up at Kaapmuiden for a longer period than is now allowed in order that he may do a bit of business.[22]

The desire on the part of white railway staff – always modestly remunerated – to slow the passage of the migrants' money through a station and benefit from it was at its most intense during the Great Depression, and nowhere more so than at Komatipoort. There, at the last stop on the South African side of the border with Mozambique, the stationmaster, unwilling to be found to be profiting directly from the platform trade, made sure that he remained on good terms with the local traders and they with him. How that friendship might have been cemented is not too difficult to imagine, and it fell to the WNLA's District Manager in Lourenço Marques to explain what lay behind the delays that bedevilled the onward connection:

> I beg to draw your attention to the recurrence of the constant unnecessary delay of our down train at Komatipoort, which, as you are now doubt aware, is caused by the Station Master wishing to oblige the local traders. For example, on the 12th inst. the train arrived at Komatipoort at 8.00 am and only reached Ressano Garcia at 9.45.[23]

Looking back up the line, then, one can see that from the moment the 307 heaved out of Booysens station, in Johannesburg, until it finally crawled across the narrow strip of land separating Komatipoort from Ressano

Garcia, formal and informal hawkers on the platforms of all the main stations down the Eastern Main Line strived to slow down the progress of the train and extract as much money as possible from the repatriates. Most of these transactions were doubtlessly perfectly legal and willingly entered into by sellers and buyers alike in confined spaces, with hurried trading and the dispensing of change being part of a race against the clock and the departing train. But there was also a good deal of platform-persuading of the men from the Sul do Save that was ethically dubious and oftentimes manifestly criminal in intent. Almost all transactions, however, took place under the cover of what, for a time at least, seemed like a 'market', one in which the skills of the vendors were pitted against those of the buyer in what, ostensibly at least, was full and fair competition, albeit often in semi-darkness.

But, within the confines of the station, on the down-train and well beyond, there often lurked other, totally hidden dangers in the form of unscrupulous or violent men intent on a more direct acquisition of the returning Mozambicans' money. For them, there was no need of the marketplace and an element of competition. These train robbers and thieves had good reason to take a close interest in the passengers on the 307.

During the 1920s and 1930s, Ressano Garcia-bound repatriates and 'rejects' not infrequently carried sums of cash that, by the standards of the day, were large and sometimes apparently quite extraordinary. In both cases, however, there were relatively straightforward ways of reconciling a cheap labour system with something approaching black working-class 'wealth'. In the earliest years of the mining industry, no workers – black or white – received compensation for incurring miners' phthisis, or, as it is now more generally known, silicosis. After the passing of the Miners' Phthisis Act in 1912, however, all miners who had contracted what had become the quintessential underground malady were entitled to compensation.[24] Mozambican miners, famous for undertaking more and longer tours of underground labour than their black South African counterparts, were therefore among the beneficiaries of the new Act. The final, often futile and derisory, payout could seem like an extremely handsome sum.

Any exceptionally large sum carried by a miner returning to the Sul do Save on the down-train, however, would seldom belong just to him. Rather, it was the collected savings of an entire group of miners still under contract and was destined for distribution among friends and family living in the same village, or in villages adjacent to the repatriate who had been called upon to assume the role of communal courier. Illiteracy and an all-too-understandable lack of trust in the office of the Portuguese Curator, let alone the post office, to guarantee the security of remittances by migrants back to the east coast, encouraged miners to entrust their savings to close friends or family, who, almost invariably, they characterised as 'brothers'.

In order to spread their financial safety nets as broadly as possible, miners in reasonable health who were leaving the Witwatersrand mines often broke their journey back to Ressano Garcia, choosing to spend a few days with fellow villagers in the compounds of the ultra-low-paying collieries of the Eastern Transvaal before rejoining the train a few days later.[25] In this way, then, the 307 was often carrying sums far greater than might simply be read off the 400–700 repatriates aboard the train at any one time. Indeed, the bank-on-wheels would often be carrying the wages of a significantly larger but unknown number of mineworkers still under contract. All of this helped make Witbank, where workers were routinely detrained for feeding at the WNLA compound, a favourite haunt for pickpockets and wide boys, as well as a starting point for a few more ambitious train robbers.[26]

By focusing on the plight of individual cash-starved miners returning east at the end of their contracts, it is all too easy to lose sight of the *collective* value of savings and wages being shipped home on the twice-weekly 307 down-train. It was a mistake that those who were intent on preying on the migrants and the train never made, and helps account for the unquantifiable volume of direct theft, and indirect theft (via forced currency exchanges at the border), that was such a marked feature of the interwar years, when the number of Mozambican labourers being sucked into the collieries and gold mines annually was expanding at a historically unprecedented rate.

In 1924, it was suggested by those in the know that each returning Mozambican miner aboard the 307 was carrying, on average, about £5 in cash between Booysens and Ressano Garcia.[27] Each departing train conveyed, on average, about 500 migrants, which meant that that every time the train left Johannesburg it was moving about £2 500, excluding any one-off settlements deriving from the miners' compensation funds, which would have raised that sum significantly.

Adjusting for inflation and converting currencies, this means that, in 2018 terms, every twice-weekly down-train out of Booysens was carrying on average around £150 000 in value, or, in current South African terms, over R2 million. The shadowy night trains, operating in semi-darkness and deliberately halted at poorly lit, unguarded stations, had entrails of pure gold.

Repatriates who tried to avoid carrying too much cash on their person, by hiding money among items of clothing in boxes stowed in the luggage van, were among the more predictable victims of theft. Cash left in jackets or trousers in hot, overcrowded and unlit compartments encouraged the plunder of the innocent and the naive.[28] Those in the hospital coaches, halfway between this world and the next, who had often been given a tot or two of brandy by the conductor to cope with their pain, were an obvious target for the depraved. 'Portuguese Native G/101684 was repatriated on or about 23 August last, received compensation amounting to 172-2-0 pounds,' the Curator noted in a letter to the WNLA, but 'When he arrived at Ressano Garcia, he only had 20 pounds in his possession and he was too ill to know what had happened to the balance.'[29] The living dead, forewarned by conductors, sometimes resisted, as when the SAR's African cleaners, equipped with carriage keys, invaded the Red Cross coaches at Waterval Boven during the Great Depression, unsuccessfully demanding that the patients hand over their money.[30]

But a cadaver mouthed no resistance at all. 'Wetele, P.P. No 25496' received 38 pounds sterling by way of compensation under the Miners' Phthisis Act in 1922 and, along with 49 other 'sick natives', was sent home in the hospital coach in which he shared a compartment with three other repatriates. His 'gamble with death', however, ended when 'he died in the train between Nelspruit and Kaapmuiden early in the morning of January 27 last'. The body was apparently not searched by the European conductor, and the police at Komatipoort reported later that, by the time they received it, there was no sign of the deceased's money.[31]

Thieves, who at first glance appeared to be just like any other men from the Sul do Save, and who dressed and talked like the repatriates, were extremely, if not impossibly, difficult to identify before they set about doing their mischief. Like certain witches, it was their very ordinariness that helped disguise them as they moved freely through the ranks of the migrants. They were especially menacing because they constituted the enemy hidden *within*. But what about the enemy *without*, those whom the repatriates were predisposed to recognise as coming from within the system precisely because they dressed differently and spoke the language of authority to the vulnerable and the weak? On the stations and on the 307, the genius of the confidence tricksters lay in their being able to command the codes of the system as a whole and pretend to be consummate *insiders* when, in fact, they were often the very opposite – ruthlessly predatory *outsiders*.

Under authoritarian rule and dominated by whites in a country where indigenous Africans, let alone black foreigners, were denied the most fundamental of rights in almost every sphere of life, the state was, to all intents and purposes, indistinguishable from a police state. For Mozambican Africans the world – at home and abroad – demanded deference: deference in the presence of the state and its many agents; deference before those public and private officials whose familiarity with the pen gave them access to the invisible power of the written word; and deference in the presence of those uniformed men – black and white – whose word was enforceable by means of batons, sticks, whips and handcuffs or the revolver. In the world inhabited by the Sul do Save migrants, there was little to distinguish clearly, or to separate, the authority of the Portuguese administration and that of the WNLA. The 804 and 307 trains were part of authority and power-in-motion commanded by hidden forces difficult to understand.

The cruder the manner in which power is exercised by those in authority, and the deeper the pattern of deference implanted in the minds of the powerless, the easier it is for the confidence trickster to mimic and manipulate the characteristics of oppression as he exercises the dark arts of fraud and theft. And who better placed to do that than those most familiar with how arbitrary power is exercised? When the confidence trickster's mischief was perfectly executed in an ethically flawed dispensation it could, like a work of art, expose the underlying fault lines of colonialism by holding a mirror to the economy and society and revealing its many pressure points.

On 3 January 1924, just as the whistle announced the imminent departure of the 307 from Witbank, 'two natives' – said to be former WNLA employees – slipped aboard from the off-side of the train carrying SAR carriage keys, a railway lamp and a suitcase. One was dressed in a 'WNLA uniform', the other in a 'sort of railway uniform'. Both 'adopted an "official" attitude', reinforcing the belief that they were 'police boys' of one sort or another. They boarded two coaches above the hospital coaches – towards the rear of the 307. Exuding confidence of a sort that suggested they knew exactly where the European conductor was at the time of boarding and later, the pair took their seats among the regular Mozambican repatriates. About ten minutes later, they got up and moved down the train, using their keys to lock the interconnecting carriage doors behind them.

Once inside the Red Cross coaches, the two men strode down the corridors, opening compartment doors and telling 'the rejects to "get their money ready"' since they would shortly be collecting all of it for safekeeping with the European conductor, and that 'they would have it returned to them at Ressano Garcia'. In a way, the suggestion mimicked aspects of the deferred-pay system that had operated briefly until the First World War, when repatriated miners were for a time paid out at the border. Returning minutes later, the man carrying the suitcase 'looked at each boy's disc number', took whatever money was handed to him by 16 of the patients, calling out the numbers to the second man, 'who appeared to be writing same down in a book'. The compliant, dulled, ill and illiterate repatriates were then handed 'blank sheets of paper as receipts'.

When questioned much later as to why they had handed over their money 'without any question or objection they simply said that they thought that it was quite alright because the police boys were Government'. With takings of nearly £200 – worth about R218 000 today – the thieves then returned to the repatriate section of the train, and sometime later, still under cover of darkness, slipped off the slow-moving 307 near Arnot. The 'rejects' and terminally ill apparently thought no more of the matter and only informed the conductor – the man supposedly in charge of the security of the train – of what had happened at around 5.00 am the following morning as the train shuddered to a halt for the usual artificially prolonged round of refreshments at the small Nelspruit station.[32]

Seen purely as *performance*, the theft was perfectly staged, displaying all the virtues of perfect timing, verbal prompts evoking the authority of the conductor and an excellent use of costume and props. Indeed, in managing to avoid being seen, let alone interacting face-to-face with the long-serving European conductor, the duo of African perpetrators displayed a degree of self-confidence and timing that would have been the envy of most thespians.

Unfortunately for the team of two or perhaps three tricksters, however, a patient in one of the Red Cross coaches had recognised one of them as a former WNLA employee who had worked at the association's Johannesburg hospital and who would have been well-known to the conductor who had organised the loading of the hospital coaches at Booysens. The suspect was eventually traced, arrested, charged, and sentenced to 12 months' imprisonment with hard labour, but his accomplice successfully eluded the police. Intriguingly, during his trial the man convicted of the crime – on one count only – chose not to question the conductor's porous written account of the theft and the public prosecutor felt no need to subpoena the conductor.[33]

The Great Train Heist of 1924 – for that is what it was – was not the only substantial theft on the Eastern Main Line that focused on stripping the men of the Sul do Save of their compensation, savings and wages. Nor was it the only occasion when imposters in 'police boy' uniforms succeeded in targeting returning migrants at stations or on the train.

Several 'confidence tricks' were reported over the following years and more especially so during the Great Depression, but in the nature of such things no doubt many more remained unrecorded.[34] There was, however, something – something never articulated fully – about the 1924 robbery that senior managers at WNLA found especially worrying. It reeked of an inside job, albeit one in which the conductor was exempted. The brazenness and scale of the theft, and the vulnerability of the sick men, made the incident exceptional. Concern was expressed that the robbery might discourage the flow of inward-bound migrants to the mines if the ill and dying miners returned to their villages totally penniless. Death was an occupational hazard for black miners and was sad in its own right, but a worker returning home to die a pauper's death was tragedy compounded. The 16 victims thus all received *ex gratia* payments from the WNLA, compensating them for their loss, but by the time the cash was handed over a few weeks later two of the victims were already dead. News of the payments quickly did the rounds among Mozambican miners, and from Witbank – the very heart of industrial darkness – it was reported by European managers that the settlement 'had created a very good impression in the native mind and had considerably enhanced the Association's reputation for fair dealing with the natives'.[35]

Maybe so, but neither the public prosecutor in the 1924 case nor the WNLA management had reasons for complacency. Indeed, had they been only slightly more mentally alert they might have wanted to probe the background to the Great Train Heist with a lot more rigour and vigour. Circumstantial evidence strongly suggests that the European conductor on the 307 on the night that the Red Cross coaches were invaded was probably acting in concert with the two confidence tricksters, whom he would have met and interacted with on several occasions – first at Booysens, and later at Witbank – over the preceding 48 months.[36] More importantly, the conductor had returned to work only a few weeks before the heist, having been previously suspended from his duties for three full months – without pay – by the WNLA for his part in a deeply exploitative, morally indefensible currency exchange scheme at Komatipoort.[37] The conductor was a man with an eye for the main chance.

Portugal's belated entry into the First World War cost the young republic dearly and served only to further exacerbate the chronic financial instability that plagued the metropole and its colonies alike. Inflation during the war halved the value of the escudo but it was in the post-war years that it spun entirely out of control: 'There had been about five escudos to the pound sterling before the war, and in 1924 there were nearly 130 to the pound, which had itself lost about half its value.'[38] For the international migrant workers of the Sul do Save this meant that, although wages for black miners in South Africa were actually still declining in real terms, their sterling-based earnings, paid out in Union notes, were steadily increasing in value across the border, where the Lisbon government had sanctioned local bank-issued currency deemed to be, in yet more wordplay, 'Portuguese sterling'. Of course, the appreciated value tucked away in Union notes did not escape the notice of the Mozambican border authorities, who proceeded to organise the official – and unofficial – plunder of all those migrants aboard the 307 down-train:

> [M]en returning to Mozambique lost on average of 4s in the pound when they changed their Union notes at the border. They were given Portuguese sterling at par despite the fact that it was being discounted at anything from 12½ to 20 per cent. The Union notes went to the Banco Nacional Ultramarino [in Lourenço Marques], but not apparently before several Portuguese officials at Ressano Garcia had taken

a share of the 4s premium, using some offensive but thorough search methods to ensure that they got all the Union notes being carried.[39]

In Lourenço Marques the cash-strapped local agent of the South African mining company, Lewis & Marks, began discounting Union notes at the uncompetitive rate of five per cent, as opposed to the prevailing rates of between 15 and 17.5 per cent, but quickly tumbled to the fact that if he set up shop upstream from the Ressano Garica border post – at Komatipoort – and obtained access to the vast majority of the miners on the down-train before even the Portuguese officials got a chance to strip them of any added value, he could make a killing by benefiting from the sheer volume of trade. Having established that he did not require a licence to engage in foreign-exchange transactions on the South African side of the border he linked up with a local Komatipoort merchant and the WNLA conductor, who, it was thought, had been offered a commission on any business entered into with the returning migrants.[40]

By early August 1923, whenever the 307 coughed its way into the station at Kaapmuiden, the helpful conductor 'advised' all the migrants to get their money ready for changing at Komatipoort, where the Lewis & Marks agent, who must have obtained permission from the notably business-friendly stationmaster to set up his table on the platform, was waiting on the incoming train. Normally, the carriage doors of the 307 were locked during the drawn-out stopover at Komatipoort prior to crawling across the border strip to Ressano Garcia, but that month the customary procedure suddenly changed. The coach doors were unlocked and – so the migrants said – they were all compelled by the conductor and the businessman and his associates to exchange their Union notes for escudos at rates even more unfavourable than those available from the Portuguese authorities further down the line.

News of a rival, upstream, compulsory foreign-exchange system operating on the South African side of the border was poorly received downstream by Portuguese officials, who reported the disruption of their own official, state-regulated trade to the authorities in Lourenço Marques. The disagreement hinged, in essence, on who would be allowed to strip out the most value from the repatriates' Union notes, and threatened to disrupt the normally cordial relations between South Africa and Mozambique. Tipped off about the conductor's role in a messy business by his District Manager in Lourenço Marques, the WNLA General Manager in Johannesburg hastily suspended the conductor. The returning migrants, having been freed from the ravages of the upstream thieves, were left to the tender mercies of those downstream. And, of course, the downstream robbers remained active.

Only a few months later, in 1924, there was more official trouble at Ressano Garcia when a number of the Asian-born civil servants employed by the Portuguese administration, too, got involved in another large-scale foreign-exchange theft from WNLA repatriates.[41] Like highwaymen in a mountain pass waiting for the passage of a stagecoach in order to prey on the passengers, so the border officials at Komatipoort/Ressano Garcia lay in wait for the incoming railway coaches to rob the returning migrant workers. It was grand larceny.

For half a century and more, the men of Sul do Save were robbed of their freedom to choose an employer by the Mozambican and South African governments even before they left the country of their birth; on the Witwatersrand itself they were robbed of the true value of their labour by mine owners intent on reducing wages for the benefit of shareholders in the developed world; and upon their return to the border post leading to their homeland the miners were robbed of the true value of their savings by border officials, train conductors and unscrupulous dealers when they were forced to exchange any holdings in 'foreign currency' at fraudulent rates.

From the instant they boarded the down-train at Booysens with their cash savings and meagre worldly goods to the moment they cleared the border post at Ressano Garcia, the returning miners of the Sul do Save – the seemingly healthy and the living dead alike – were seen by many, if not most, in South Africa, black and white alike, as little more than objects of prey. Like the odious border officials, the non-state platform and station predators knew exactly where to lie in wait for the repatriates. The iron road home, which seemed to promise a relatively safe and swift journey back to a familiar rural world, was more gauntlet than gateway.

And, while it was true that for those 'East Coast Boys' not travelling on to Lourenço Marques by rail, or onward by ship to Inhambane, the 307 gauntlet might have appeared to end at Ressano Garcia, in fact it never did. In 1916 – but it could just as easily have been in 1926, 1936, 1946 or 1956, and just as true – an observant Anglican missionary noted of the unfortunate repatriates:

> Whenever they land, or leave the train, there is someone on the *qui vive* to make something out of them, a drink shop keeper, an official, a low 'white' pretending to some authority, a low woman, in places even footpads who attack and rob by force. Friends most unexpectedly appear to ask for a loan or to share in some expenditure and when home is reached there is the customary gift to be paid to the chief – probably a feast to be given.[42]

The promise of the civilised world of hard work as the road that led unfailingly to progress and prosperity sometimes rang hollow for the hundreds of thousands of black Mozambicans locked into the greater southern Africa economy. In much the same way, families in the Sul do Save learned, over a half-century of pain, that the much-vaunted benefits of modern technology – as supposedly embodied in the railways – were never quite hazard-free, and nor could their fatal shortcomings always be easily separated from colonialism, gross exploitation and racism. Modern technology is never class-neutral; the privileged or the wealthy, sometimes both, public and private alike, are always better placed to acquire it early and usually benefit most from it over the longer run. What did the railway to the Rand ever do for ordinary men and women in southern Mozambique?

# PART IV

# CHAPTER NINE

# WNLA Train Disasters

## Categorisation and Causation in a Colonial Dispensation

*Accident is the name one gives to the coincidence of events, of which one does not know the causation. Accidents exist only in our heads, in our limited perceptions.*

— FRANZ KAFKA

An act of God can be defined as 'an instance of uncontrollable natural forces in operation'. An accident, by contrast, is not necessarily locked into the matrix of natural forces. It is 'an unfortunate event that happens unexpectedly and unintentionally' or an 'event that happens by chance or that is without apparent or deliberate cause'. Within Aristotelian thought, an accident is 'a property of a thing which is not essential to its nature'. Lewis Hyde, following Aristotle's lead on *Categories*, expands on the latter in a helpful way with the following observation: 'Accidentals are present by chance, essentials by design. Accidentals are changeable and shifting: essentials are stable.' And so, concludes Hyde, 'the real significance of a thing lies with its essences, not with its accidents.'[1] In our search for the causality behind WNLA train disasters, then, it is important to distinguish between the essence of a thing that flows from design and the accidentals that occur randomly within a system.

The Eastern Main Line was a cross-border, international railway system linking two colonial regimes that, while differing in operational terms in certain, sometimes important respects, shared essentialist, racist thinking insofar as both deemed the majority of their black populations to be inherently inferior and therefore not automatically entitled to aspire to full citizenship with all its attendant rights under law. If South Africa and Mozambique differed in that regard, then it was always a matter of degree more than of kind. Their racist ideologies, whether formally adopted and explicitly articulated as state policy or inferred informally and acted upon by those privileged by virtue of the colour of their skin, informed almost all aspects of the everyday functioning of the economy, polity and society, and perhaps nowhere more so than in the operation of their railway systems.

Deeply imbricated in state machinery and inherently quasi-militaristic in their operational realities, railway systems almost invariably reveal the economic priorities of governments, as well as the real and hidden social structures of class and/or colour through the manner in which coaches and passenger travel are organised and the rules governing their use enforced. Such systems are not the products of chance; they have a shape and purpose dictated by design that reflects ideology clearly.

Here, we might pause to note two incidents in early modern South African history that illustrate how when class, colour, legal training and social status collided with railway realities they could have a major, if not seminal, impact on the developing political consciousness of people of colour. It is well known that the seeds of Mohandas Gandhi's political consciousness and activism were planted and began to germinate when, on a winter's night in May 1893, the 24-year-old barrister, a product of London's Inner Temple, was ejected from a Pretoria-bound train and left stranded on the station platform at Pietermaritzburg because a white man had objected to his presence in a first-class carriage.

Nearly two decades later, in August 1911, at a meeting of the South African Native Convention – a forerunner of the African National Congress – a motion was passed unanimously deploring 'the poor treatment of black people on trains'. Barely four weeks later, in September, one of the delegates to that meeting, another barrister, this time from the Middle Temple, London, and the driving force behind the formation of the African National Congress in 1912, Pixley ka Isaka Seme, was involved in another racial altercation while aboard a Volksrust-bound train. It was an experience that his brother lawyer, Gandhi, would have recognised all too readily from his own traumatic experience.

Seme claimed proudly that, like all solicitors, 'I of course travel first class' and requested that the steward provide him with bedding for the night-long travel. When his request was ignored he went in search of the steward, whom he found in the dining car, in the presence of five young white louts all the worse for liquor. The black lawyer's insistence on being served produced outrage among the drunks, who advanced on him, urging that he be thrown out of the train window. Seme responded by shifting the revolver he was carrying in his hip pocket to a side pocket in his jacket – a movement that produced a flash of sobriety in the young men but one that also left the scar of resentment. Two weeks later, Seme, charged with pointing a firearm at a group of people, was found guilty and convicted of the offence in the Heidelberg magistrate's court.[2]

It might not be too much of a stretch, then, to suggest that, in colonial South Africa, racism experienced aboard trains by the eminent and the educated helped shape subsequent movements of political resistance by Africans and Indians alike. At the other end of the class and race spectrum, however, nothing could be more enduring or more stark than the difference between a first- and third-class rail ticket when it came to ordinary African commuters and travellers. In colonial societies the railway systems, more often than not, laid bare the grid underpinning the political economy of the larger society – in Braille.

The 804 and 307 trains, operated privately by the WNLA on publicly owned tracks, embodied much of what was crass, exploitative, prejudicial and unethical in the cross-border trade in cheap black labour across southern Africa for the first half of the 20th century. And it was precisely because the system, taken as whole, was so profoundly racist in nature that managers and operatives alike in the SAR and WNLA struggled to settle upon the mutually agreed categories best suited to the managerial ideologies and practices that might facilitate their day-to-day operations. Where, then, did the essence of the problem lie?

Were blacks 'passengers', fully human and deserving of the same facilities and treatment as white passengers, or were they perhaps closer to animals or, worse still, mere objects – just things? If Africans were fully human, then both the 804 and 307 qualified as passenger trains that had to pass through the system as expeditiously as possible, but if they were not, then trains conveying black labour to the mines were, at best, classed as 'mixed' trains or, at worst to be left undesignated and treated as just so many goods or freight trains.

We have already noted how these problematic categories bedevilled the everyday operating efficiencies of the WNLA trains, but what we need to do now is to establish how these same classificatory problems – rooted in racial thinking – might have contributed to train accidents on the Eastern Main Line. Acts of God were unavoidable disasters, and the down-train was never exempt from them, and they can be illustrated readily enough. But there were other tragedies, 'accidents', stemming from managerial and ideological confusions rooted in the thinking that characterised the operation of the railway system as a whole, and the WNLA trains in particular, that made them acts of men alone.

The problem had deep historical roots. God's often mysterious acts in the mountainous district around Waterval Boven started almost unnoticed with the customary commencement of the summer rainfall season late in 1917. By early January, however, it was clear that rainfall that year was going to be much higher than usual, and it persisted, leaving the moody Elands River in spate for some time thereafter. The weather on 14 February 1918 was changeable – light rain alternating with heavy showers all day – and it was still overcast when the 307 pushed into Waterval Boven station shortly after midnight, on the morning of 15 February. The train comprised 14 third-class carriages, with two hospital coaches at the rear, although, in truth, it was to be another five years before these were replaced by the coaches of a later generation. In keeping with the times and the industrial war being waged underground on the Witwatersrand, the train was jam-packed with evacuees and the walking wounded. Of the 646 miners from the Sul do Save aboard the 307, no fewer than 102 – that is, about one in six – were 'sick boys' ostensibly in the sole care of the European WNLA conductor.[3] Not all of them would have been accommodated in the ageing hospital coaches.

The tracks between Waterval Boven and Waterval Onder had been subjected to routine checks by the SAR maintenance staff only hours earlier and declared safe. After the usual bouts of locomotive-switching and irritating raids on the temporarily stranded passenger coaches by invading hawkers, the train puffed its way out of the station at 1.57 am, set for the slow and sleepy glide down into Nelspruit. Eight kilometres beyond Waterval Boven there was a gradual curve to the track at which point, as the engine driver knew, the route immediately gave way to Ferguson's Cutting, where the river lay below on the one side of the track and a man-made cliff on the other. The headlight on the locomotive picked out an obstruction possibly two metres high in the cutting ahead so the driver 'immediately applied the automatic vacuum brake and put steam against

169

her'.[4] But even at the restricted regulation speed of 14 miles an hour over that particular section, it was too late. The cowcatcher plunged straight into a heavy, greasy mixture of boulders and earth. The momentum of the train was so powerful that it pushed the locomotive and the coach immediately behind it – coach number one – further into the detritus lying across the rails, leaving them partially buried beneath a landslide caused by the persistent heavy rain.

The driver dismounted, checked that the wheels of the locomotive and the trailing coaches were still squarely on the rails, and concluded that, even so, he would be unable to proceed until such time as the debris had been cleared from the track. At that point, the WNLA conductor, having already checked on the condition of the patients in the two hospital coaches, appeared at the front of the train to satisfy himself that the passengers in coach number one were unharmed. When he established that they appeared to be in no immediate danger, he promptly returned to the rear of the train. But no sooner had he left than another boulder came crashing down the cliff face, landing on the roof of coach number one. This so alarmed the driver that he at once went to the rear of the train and suggested to the conductor that all repatriates in the first coach needed to be evacuated immediately in case there was another landslide, which seemed as imminent as it was likely.

The driver immediately returned to the front of the train and waited on the conductor to reappear to evacuate the passengers in the front coaches he considered to be at greatest risk. He and other railway colleagues said later that it was quite a wait for the conductor to return. What next happened lay squarely within the domain of man rather than God and became the subject of dispute between the driver and the conductor that was never resolved by the Board of Enquiry called to determine the cause of the tragedy that followed. It was not clear how long it took the

conductor to reappear at the front of the train, but before the repatriates could be removed safely from the front coach there was a second,
more substantial landslide, 'falling on the stationary train and completely
crushing in the centre of the second coach from the engine, No. 3030.
There were some 40 "natives" in the coach and 11 were killed, presumably
instantly, one seriously injured, and nine others slightly injured.'[5]

What was at issue, then, was the question as to whether or not there
had been sufficient time to remove the passengers from *both* the front
coaches before the second landslide occurred. Had there been an avoid-
able delay that had fatally compromised the lives of the repatriates in the
second coach? An act of God had undoubtedly given rise to both the
first and the second landslides, but was the crushing of the second coach
entirely unforeseen? Was it merely an 'unfortunate event' that 'happened
unexpectedly'? Did what happened at Waterval Boven in 1918 qualify
as an 'accident' or could it perhaps have been attributed to some more
deep-seated cause? Could equivocation by the conductor as to the precise
nature and status of the black passengers – a design-essence problem in
the system – have contributed to the 'accident' that followed? In short,
could racist thinking at every level of the greater operating system, part of
an ongoing classificatory problem when it came to operating the trains,
have been a contributory *cause* of the accident?

The driver of the locomotive on the 307 could not have been clearer in
his evidence to the Board:

> I think that the conductor had sufficient time to get the boys out of
> the coach if he had taken my warning, and I do not think that there
> was any risk involved. I think it was possible for them to get out of
> the coach on the lower side and walk along the line. I have walked it
> several times and I think it is safe.[6]

The conductor's version recalled the decisive moments differently:

> After the driver spoke to me I went to the engine at once to see
> whether I could get them out. We were going to start to get them
> out on the river side but I had only just arrived when the second fall
> occurred. I followed the driver up to the engine and arrived there
> shortly after he did … If the other witnesses state that from the time
> the engine driver told me to look after the safety of the natives that
> it was at least 20 to 30 minutes before the big crash took place it is
> not correct. It would not have taken me that time to get up to the
> engine. You must realise that in a time like this one cannot judge the
> time very much.[7]

From his own evidence, however, it is clear that the conductor had done
quite a lot of thinking prior to making his way to the engine. He was concerned that he might not be able to find an easy and safe path along which
to evacuate the passengers from the front coach. He, personally, was 'not
flustered' before he reached the engine, he claimed, and the repatriates in
the front coach were also calm: 'It was only when the big crash occurred
that they got to the excited stage.' Yet, despite that, he was clearly not
certain how the passengers might respond to any attempt to get them
clear of the coaches. 'The natives are difficult to deal with at a time like
this – half-asleep – they would be like a lot of sheep. In fact it was very
difficult when we got them out to prevent them going over the bank.'[8] The
latter evidence was never corroborated, and, taking into account that the
repatriates were for the most part hardened African miners accustomed
to negotiating far more treacherous situations deep beneath the earth's
surface, they were most unlikely to have been scared or panicked. Indeed,
they had been notably 'calm' when the locomotive first struck the debris.

The Board of Enquiry, however, chose not to pronounce on the difference between the driver and the conductor's versions as to the timing of a possible successful evacuation of passengers from the front coach, let alone the second coach, and the subsequent, much larger landslide. Instead, the findings merely recorded that:

> The driver feared that there might be another fall of rock and earth and warned the conductor of the natives, suggesting their removal from what he considered to be the threatened coach – viz. No. 1. The conductor's evidence, however, is that although the suggestion appealed to him there was no time to carry it out before the main fall occurred.[9]

The principal finding was, therefore, that the tragedy was a classic misadventure – 'an event that happens by chance or that is without apparent or deliberate cause' – and that 'The cause of the accident can only be ascribed to the effect of abnormal weather, which effects there were no reasonable grounds to anticipate'.[10] There was, of course, a measure of truth to that skewed tautological analysis but it was not entirely convincing.

The national political dispensation, decided at the birth of the Union eight years earlier, when Africans were denied even the prospect of gaining the franchise, was racist. The SAR, an instrument of the state, was, as with the political system, predicated on divisions by race. The WNLA train, beyond the reach of even the state in that it was a private initiative running along public tracks, was organised along explicitly racial lines and its conductor was unambiguously racist in his thinking and actions at the time of the accident. There was, then, a hidden chain of causality predicated on unacknowledged racism that could be traced from the functioning of the system as a whole – which gave preference to white passengers – down to the accident or, working backwards, from the accident back up to the racial construction of the state and the railway system. How best, then, to reconcile the evidence of the locomotive driver and the observations of the conductor? The answer lay in focusing only on the antecedents of the 'accident' and blaming God.

The WNLA management, pleased to accept that 'no blame can be attached to anyone for the accident, that it was an "act of God"', was nevertheless inclined to take the view that it would be only fair to 'make compassionate grants to the persons affected'. One can but speculate why that might have been so. But the Association – concerned that it might be setting a precedent – did not want to be taken advantage of by the Mozambican families. It wanted to be reassured that it was dealing only with cases of 'genuine hardship', and that 'it should be made very clear to the natives affected that any payments made by the Association [are] compassionate grants and in no way compensation legally due to them'.[11]

It is difficult to come away from the varying accounts of the events leading up to the 1918 Waterval Boven tragedy entirely free from the impression that, somewhere between the two landslides that occurred in Ferguson's Cutting that morning, there was something in the design, in the very essence, of the way the 307 train was operated, and its black passengers perceived and treated, that contributed significantly, albeit indirectly, to an 'act of God' that cost 11 men their lives in what was then written off as an 'accident'. But, should there be any residual doubt that a colonial ideology that fully embraced racist thinking might have contributed to how the railway system was constructed and run, or that it might have helped account for an 'accident' and the deaths of African passengers, then an examination of the disaster that befell the down-train in 1949 – by then operating under a new number, 513 – might help to clarify our thinking further.

The fact that the second major accident to befall the down-train took place little more than a year after the National Party, intent on furthering the policy of apartheid, won the election of May 1948 should not, however, colour our interpretation of the 1949 disaster unduly. The explicit and implicitly racist thinking that underpinned the operation of the private WNLA trains on the publicly controlled SAR system, as we have repeatedly noted, long predated the formal adoption of apartheid policies by the South African government. In fact, the underlying dynamic that informed the running of the 307 and 804 (513) was so deeply ingrained in the thinking of the Chamber of Mines and the white ruling class that the system – insofar as it related to the running of WNLA trains – remained essentially unchanged from the time of Lord Milner, in 1902, until that of DF Malan, the Union prime minister in 1948, and beyond. Politics changed over that time but the railways hardly so. Racial classification meant that, when it came to black lives and the railways, collective, formal, institutional compassion – private and public alike – remained at a premium.

The 1949 *Report of the Board of Management* of the WNLA devoted just two sentences to the operational disaster that had marred the year. 'On the 15th November, 1949, a train carrying East Coast Natives on repatriation to their homes in Portuguese Territory was derailed at Waterval Boven in the Transvaal with serious loss of life. The accident formed the subject of an official Board of Enquiry.'[12] The WNLA report contained no details as to the number of fatalities, casualties or the possible causes of the accident. Some latter-day historians interested in heritage have ignored the disaster entirely, yet it remains eminently worthy of closer examination.

On 14 November 1949, the 14-coach 513 down-train got off to a good start. Being the Monday rather than the Thursday train, which always carried a larger number of medical evacuees, it left Booysens with 404 repatriates on board, only 9 of whom were 'sick rejects' confined to a hospital coach. At Witbank the 513 was joined by another 66 time-expired returnees and, unusually, only one more 'sick boy'. As always, nobody was absolutely certain as to the precise number of people on the train since it was sometimes boarded illegally by a few eastbound free-riders, but the mood among most of the Sul do Save migrants aboard was buoyant. It was summer and not only did the fit and apparently healthy greatly out-number the living dead and the few insane, but the Mozambicans were also heading home to their families and Christmas was beckoning.[13]

Good fortune seemed as if it were intent on following the train and its passengers all the way to the border when it left Witbank promptly at 7.28 pm. There were six scheduled stops between Witbank and Waterval Boven, at which points the 513, a 'Mixed (Natives) Train' (since it might be called upon to pick up a few coal trucks along the way) had to allow other freight trains or the regular European mail train to pass. The time budgeted for the six officially planned stops was 53 minutes, but that night the locomotive driver got lucky. There were no priority passing trains, with the result that the down-train gained momentum, and by the time it reached Waterval Boven, at 11.15 pm, it was a full hour ahead of schedule. But, if good luck had somehow crept aboard the down-train, it slipped unseen into a cloudy night shortly after midnight, when the calendar ticked over to Tuesday 15 November 1949.

The Witbank crew, satisfied with a night's work well done, signed off duty, but when the incoming driver signed on, he heard of a problem down the line that needed addressing before he would be allowed to pro-ceed. A freight train was stranded at Nelspruit, waiting for a locomotive to haul it back up through the mountains to Waterval Boven. The 'running foreman' at Waterval Boven decided on a solution that often worked with goods trains but had never before been tried with a passenger train: they would add an extra locomotive to the 513, turning it into a 'double-header' for the slide down into the Lowveld. There, the extra locomotive would be uncoupled at Nelspruit and the 513 would simply proceed, as usual, to the twinned border posts at Komatipoort and Ressano Garcia.

The plan was straightforward enough, but the down-train would then require two engine drivers rather than one. That, too, was not an insurmountable problem. There were two drivers on standby: the more experienced, a 'special class' operative qualified to work passenger trains who had already signed on for duty, and another, less experienced, 'second class' driver familiar with working freight and 'mixed trains' but not passenger trains. Having a passenger train being worked by a double-header would be a first out of Waterval Boven, but the usual convention would apply: the more experienced 'special class' driver would take charge of the lead locomotive and the 'second class' driver would take charge of the second, following locomotive.

But, when the two drivers met, the 'special class' driver – for reasons that were never uncovered – indicated that he did not want to take charge of the lead locomotive, which meant that the 'second class' driver was 'forced into a position for which he had insufficient training and experience'.[14] In the mind of the 'second class' driver – now the lead driver – this reversal of roles might have been further complicated by the fact that it was not clear what exactly the status of the 513 was. Formally, it was a 'Mixed (Native) Train', a long-standing, troublesome category insofar as it most certainly was not a conventional 'passenger train', in the sense that it transported whites, even though the WNLA had, for decades, for reasons more industrial-logistical than humanitarian, repeatedly pleaded with the SAR to treat it as such. In operational terms, a 'mixed' down-train was more of a half-and-half conveyance – half-goods and half-passenger. For all that, the 513 looked and moved like a passenger train, even if its passengers clearly were not only 'natives' but 'foreign natives' to boot. In truth, the train had always been – and still was in year two of apartheid – almost unclassifiable in the minds of the government, railway administrators and members of the white South African electorate.

The 'Enquiry Board' that examined in great depth and with considerable professionalism what transpired later was fully aware of how such cognitive confusion might have contributed to a fatal operational error:

> As [the 'second class' driver] had had a considerable amount of experience in the driving of goods trains over this section, but none in the driving of passenger trains, he may have had the thought in his mind that passenger trains were required to be hauled at a considerably higher speed than goods trains …[15]

If that were true, then the 'second class' driver, pushed into an unfamiliar position by a senior colleague, would not have had his troubled mind eased by the fact that, during the stopover at Waterval Boven, the train took on a further 18 time-expired workers, or that the marshalling of the double-headed train took up more time than had been scheduled.[16] With its lead locomotive now being worked by the unqualified 'second class' driver, the 513 left Waterval Boven eight minutes later than scheduled and headed straight for the gradually accelerating, but speed-restricted, descent through the Devil's Throat and towards the bridge across the nearby turbulent Elands River.

The experts serving on the Board of Enquiry found it impossible to determine the exact time that the accident took place. In part, that was due to implausible statements made by certain SAR staff members who, keen to shield their colleagues, offered times that could not readily be reconciled with the scientific calculations relating to the tragedy. What was

*The scene of the 1949 train accident at the Elands River bridge, near Waterval Onder.*

clear, however, was that the train picked up speed almost from the moment that it left the station, and that about five minutes out of Waterval Boven, as the double-header got to the permanent 22-miles-per-hour speed-restricted section leading into the gentle bend and curved bridge over the Elands, the train was already travelling at about 40 miles per hour. The double-header sprinted across the bridge, but, as it approached the far side, the speed of the train shifted the weight and centre of gravity of the twinned locomotives and their tenders, which capsized. As they toppled over, 13 out of the 14 saloon coaches flipped off the rails, with 6 hurtling directly down into the Elands about 21 metres below.[17] Ironically, only the hindmost – Red Cross – coach remained firmly on the tracks. Never were the limitations of the largely cosmetic hospital coaches more cruelly exposed.

Sixty-two sons of the Sul do Save lost their lives on impact, 106 were seriously injured, and many more suffered minor injuries. In a further irony, the 'special class' driver was killed, while the 'second class' lead driver was left severely injured. When news of the disaster reached Waterval Boven via a rail-side telephone some minutes later, 'the warning siren at the power station was sounded continuously for some time and this had the effect of rousing most of the townspeople, many of whom, when they became aware of what had happened, proceeded to the scene to assist'. Included in that number were 'a number of members of the native branch of the St John's Ambulance Brigade, Waterval Boven'.[18] The recovery of the bodies and rescue of the injured and other victims were made all the more remarkable because they took place largely in the darkness and gentle rain.

An ambulance train 'consisting of third-class saloons, a truck, a dining car and a first-class saloon', accompanied by medical doctors, left Waterval Boven at 11.50 am. The first stop was at Belfast, where the Union Defence Force had made available three Dakota aircraft, which transported 40 of the most serious cases directly to Pretoria for treatment at the general and military hospitals in the city. The remaining victims were accommodated in hospitals at Middelburg and Witbank.[19] As had happened during the 1918 accident, the WNLA's hospital coach emerged from the tragedy fully intact, but in terms of the possible service that might be rendered to the injured from it, or from within its so-called wards, it was hollow, perfectly useless. Nor was that the only indictment of the cross-border rail system as operated by the mining industry. Six of the victims could not be identified and, to this day, are listed simply as 'unknown' on a memorial stone erected above a mass grave in what once was a perfectly racially segregated quarter of Waterval Boven.[20]

The Board of Enquiry investigating the disaster was fully aware that what had happened at Waterval Boven in November 1949 could not be attributed to chance, and that it therefore did not, strictly speaking, qualify as an 'accident'. Indeed, the root causes could, in good part, be traced to some of the conceptual confusion that arose from the racial thinking that informed the categorisation of trains carrying African passengers along the Eastern Main Line and other parts of the extended South African system. The Board made five recommendations arising from its inquiry into the tragedy, two of which were interlinked. The first underscored the need for absolutely rigorous enforcement of regulations governing the qualifications of those permitted to drive passenger and goods trains, but it was the second recommendation that went to the very heart of the matter. It is worth quoting in full because it laid bare the culpability arising from racial thinking:

> In view of the apparent difficulty and confusion which arises by designating train No 513 as a 'Mixed (Natives)' when, in fact, it is a passenger train, consideration should be given to amending its designation so as to leave no doubt in the minds of the staff as to what is intended, or, if this is impracticable, suitable instructions should be issued clarifying the position. This recommendation might well be examined, not only so far as train No 513 on the Eastern Transvaal System is concerned, but in respect of other Systems where trains might be similarly designated.[21]

And so it was that, in year three of apartheid – 1950 – the South African Railways administration, for the first time in its 40 years of existence, attempted to grapple with the ongoing problem as to whether black men – 'natives' – travelling on trains might actually be designated as passengers, albeit of a special sort, rather than as part of some other, composite category such as 'human freight' that might have contributed to operational difficulties and fatal 'accidents'. In that same year, the SAR still reported to Parliament on the WNLA trains under the old subheading of 'Natives by Goods Trains and Special Native Trains'. But the Waterval Boven accident may have been a turning point in that, in the eyes of senior railway administrators at least, the words 'human' and 'freight', as used by SAR and some lowly WNLA operatives, had become uncoupled and the trains formally redesignated. By year four of apartheid – 1951 – any reference coupling the words 'Natives' and 'Goods Trains' had been excised from the SAR report to Parliament, and the old 307 and 804 trains were included under the new, euphemistic heading of 'Special Native Fares'. Africans, including the Mozambican miners, were no longer to be bundled up into the same category as batches or freight, but nor were they just 'passengers'; they were humans, but humans of a special kind, 'natives', black men who, as the cancer of apartheid swiftly spread, would soon be reclassified as 'Bantu'.

For half a century, African miners on the up-train were never referred to as 'miners' but as 'East Coast Boys'; the living dead and terminally ill Mozambicans on the down-train were never referred to as 'patients' but as 'rejects' or 'sick boys'; and the Sul do Save migrants never qualified as 'passengers' but were 'special fares' or 'Bantu'. Racism, a product of the mind, perverts the thinking of the oppressors and the oppressed alike.

It was ironic, then, that it was under the apartheid regime that, as far as the railways were concerned, black passengers were lifted clear of the animal or goods categories while travelling on trains and their essence as human beings reluctantly and indirectly conceded. But the badge of colour, present from the beginning of colonialism, could never be eradicated and remained the primary basis for separation in every possible sphere of social life in the country. It was one – minuscule – step forward for classificatory man, and two giant steps backward for mankind as a whole.

The political and social legacy of the Waterval Boven disaster was deeply ambiguous.

In racist dispensations, the manner in which railway administrations utilise categories – and then act upon them – can never be fully separated from a deeper understanding of disaster or tragedy. Lewis Hyde perhaps says it best: 'The intelligence that takes accidents seriously is a constant threat to essences, for in the economy of categories, whenever the value of accident changes, so, too, does the value of essence.'[22]

# Conclusion

*Railroad, n. The chief of many mechanical devices enabling us to get away from where we are to where we are no better off.*
— AMBROSE BIERCE

Over the four decades under review in this short excursion, the men of the Sul do Save, their families and homes firmly rooted in a political economy even more exploitative and hostile than that of South Africa, never ceased expressing their distaste for the Mzilikazi system as a whole, or the manner in which the WNLA trains were operated. Hundreds of thousands of Mozambican families simply fled the forced-labour regime on the east coast for South Africa, leaving the Portuguese authorities, operating what was a robber economy, dismayed by 'clandestine emigration' but largely accepting of the physical ruination of generations of African lives that had been sold into servitude in exchange for a guarantee of rail traffic to the port of Lourenço Marques.

At the South African points of production – the roundly detested Eastern Transvaal collieries and the slightly better tolerated Witwatersrand gold mines – Mozambican miners offered such resistance as was possible in a system that, for the vast majority of black workers, was effectively a labour market controlled by a police state. Prompted by an extraordinary decline in the real value of wages during the First World War and the years immediately thereafter, Mozambican miners – more vulnerable than most – manifested a degree of political consciousness that, at very least, matched that of other similarly oppressed black workers. And, in attempts to rouse their fellow countrymen, the miners of the Sul do Save sometimes made use of the railway network and stations as points of reference for men corralled in mine compounds backed by fingerprint and pass systems. Some of this can be illustrated from a handbill that circulated in Johannesburg in 1919:[1]

## THE MEETING OF EAST COAST NATIVES
### Union of Natives of the Mozambique Province

Now then Gazas, Tongas, Inhambaan, Sabie, Maputu, you who under the Portuguese Flag, this is your time that you may meet at old Jumpers G.M. Co., Cleveland, in the Jews Hall on Sunday, 11th May 1919, at 10 a.m. Remember the Station is 'Cleveland'.

Let us be united as the other nations.
The saying is 'No fowl assists to feed another' or in English 'Each man to himself, God stands alone'.
Don't miss
Yours,

| D.C. Mozwoya | Chairman |
| D. Masoboni | Secretary |
| G.E. Mlungu | The Owner |
| J.T. Ngonyama | Assistants |
| G.S. Khosa | |

A similar pattern of disciplined, thoughtful and peaceful pan-tribal organisation was evident in a wave of strikes involving hundreds of 'East Coast Boys' on the Rand at the Ferreira Deep Mine, and right across the collieries in late 1918, including the Tweefontein, Clydesdale, Blackhill, Anglo-French Blackhill and Landau coal mines. The disputes revolved around the issues of contract length, proper payment for work duly done and, almost always, food rations.[2] It may be a delicious irony that, by collectively terming them 'East Coast Boys', the WNLA itself helped foster an enhanced sense of unity and heightened political consciousness among southern Mozambican workers drawn from what otherwise were very different ethnic groups.

The wartime and post-war strikes were met by a largely predictable set of responses from mine managers and the South African Police, who marched in lockstep when confronted by militant black workers, let alone less-well-understood 'foreign natives'. Invariably cast as 'outside agitators' rather than mine-based 'strike leaders', as men who led 'native unrest' rather than 'industrial action', the most prominent activists were systematically rooted out and blacklisted, which ensured that they would never again work in the Union. In one instance, the mass of workers was forcibly driven back down the mine shaft and barricaded in until the situation returned to what passed for normal. An industry that was organised in a singularly abnormal way encouraged regular and widespread informal resistance through the faking of illness, go-slows, the destruction of mine property and the theft of equipment.[3]

Perhaps inevitably, all these manifestations of formal or informal resistance by the more politically conscious strike leaders from the Sul do Save culminated in yet another final train ride back to Ressano Garcia and into the arms of the waiting Portuguese administration. During the last quarter of 1918 WNLA conductors, already accustomed to incarcerating 'recruits' on the up-journey by locking the coaches, or confining the manifestly insane to sealed compartments on the down-journey, took on additional custodial duties. They escorted strike leaders who, after deportation, were subjected to an additional 'state within a state' sentence when the WNLA made sure that they would never again be allowed to return to work in the mining industry.[4]

All of this should be seen within a wider context. Southern Mozambican men had an acutely well-developed understanding of the regional political economy as a whole, as well as what the potential cash value of their labour might be at various historical junctures. Indeed, so sharp was their insight that, for the better part of three decades, the coaches and trucks attached to the WNLA up-train had to be locked down in order to prevent desertions from the 307 as it passed through the various stations of the greater Witwatersrand. For the most adventurous of these young 'recruits' – some having been snared in forced-labour nets spread across the Sul do Save by the tax-hungry Portuguese authorities – the train from Ressano Garcia to Johannesburg made for little more than a free ride into the heartland of industrialising South Africa. Tens of thousands of them never returned and were forever lost to Mozambique. But not even those relatively resigned to having being trapped in a transnational deal made at their expense by two colonial powers went to their underground destinations mute or without protesting. WNLA train conductors, who had to live with the vagaries of the mass-transit indentured-labour system on a daily basis, felt sufficiently menaced by ongoing mutterings of varied intensity to warn regularly about the dangers of 'demonstrations' or, more darkly, harking back to the era of African armed resistance in the late 19th century, about the 'natives becoming restless'.

Hardship on the train up and the train down was part of a liminal experience – a corporeal ordeal extended over time and through space

– central to the Mozambican miners' experience of the circuit of mine labour.[5] The train, *kulukane,* said the Thonga, carried everything – men, freight and goods alike – and swept all along with it. Nobody emerged from the round trip unchanged if not unscathed. Those who had completed their spell of migrant labour on the mines successfully – the *gayisa* – returned home not only as relatively moneyed folk but also as men who enjoyed a new and elevated status in what otherwise were fairly closed and isolated rural communities. In important ways, mine labour, the journey and the wages earned opened up the chance for a fair number of men to express various new forms of informal learning and the opportunity for limited social mobility. In a severely circumscribed economy, many men from the Sul do Save therefore signed on for consecutive contracts on the 'popular' Witwatersrand mines.

In terms of both the value and volume of repatriated earnings in a currency that simply dwarfed the escudo, the men of the Sul do Save, over half a century, helped to transform the forbidding natural terrain of southern Mozambique into one large, fertile tax estate for the cash-starved Portuguese administration in Lourenço Marques. But, a colonising power more interested in maintaining a predatory tax regime – a robber economy of diminishing returns given the long-term decline in mine wages – than in forms of sustainable development, simply allowed most of the flow of cash into the region to drain away in the sandy soils of the coastal lowlands. Largely content to preside over a rudimentary commercial economy feeding off an interconnected network of rural trading stores run largely by Asian immigrants, who longed to return to their lives and families in India, the colonial administration gave almost nothing back to the Sul do Save by way of the development of education, health, infra-structural or welfare facilities in proportion to the taxes that it harvested. In terms of 'trickle-down' economics the region was a disaster, and today, with 70 per cent of the rural population still living in abject poverty, the country as a whole is deemed by various United Nations agencies to be among the ten poorest on earth.

The developing socio-economic realities in southern Mozambique, and the equivocal role played by the Eastern Main Line in helping to bring them about, were, however, never seriously questioned either in Pretoria or Lourenço Marques. On the contrary, the administrations on either side of the border – the one in awe of the Chamber of Mines and the other feeding off the principal railway artery leading into the Witwatersrand heartland – saw nothing in the Eastern Main Line but great white achievement. This unwillingness to acknowledge the dark side of the railway, or its cost in human terms within the Sul do Save or on the Witwatersrand, was much in evidence when, in 1945, new station buildings were unveiled at Komatipoort and Ressano Garcia as part of the 50th anniversary of the official opening of the Eastern Main Line. It was an occasion for much official back-slapping and triumphalism of a sort arguably more suited to 1845 than 1945.

The South African prime minister, General JC Smuts, nicknamed 'Slim Jannie' (Smart Jannie), was cagey enough to dip only his big toe into the pond of clichés available on such occasions. He wished to see 'the continuance of the close cooperation politically, economically and culturally' that had always existed between the two countries. His Minister of Transport, FC Sturrock, however, was willing to go in a wee bit further and affirm not only that the Eastern Main Line had proved to be 'mutually advantageous through the years' but that he was satisfied that the cooperative venture would 'continue to endure' (*sic*). The Governor-General of Mozambique, João de Tristão de Bettencourt, however, was willing to wade so far in that he was in danger of his words losing all meaning: 'The railway has proved to be a powerful instrument in the common task of advancing civilisation and economic expansion for the prosperity of the Transvaal territory and our principal seaport.'[6] It was the old colonial dream again, 'civilisation' and 'economic progress' marching in lockstep but with the long-term fate of the colonised left largely undefined.

The Second World War had ended barely eight weeks before these sentiments linking civilisation and the railway in so unproblematic a fashion were paraded along the border between Mozambique and South Africa. But anyone with an interest in history, let alone someone who had served in the British War Cabinet, such as Smuts, who should have borne at least a passing knowledge of how the WNLA trains had operated for fully half a century, might have chosen to have moderated the prevailing triumphalist tone with a cautionary word or two. The operation of the transnational migrant-labour service between Ressano Garcia and Johannesburg had, for five decades, carried with it a faint warning as to how, when racist thinking, steam technology and political power were brought into alignment, they carried within them the potential for crimes against humanity and, in its most extreme form, genocide.

Before 1914, German capital, in the shape of Deutsche Bank and others, was heavily invested in the construction of a railway planned to run all the way from Berlin and through the Ottoman Empire to Baghdad, even though it was not until 1940 that the first train from Istanbul reached Baghdad. The Ottomans placed their railway department in the charge of a German official, Lieutenant Colonel Böttrich, who made certain that all his policies were in line with Ottoman government thinking while many of the day-to-day operations along the line itself fell into the hands of an Armenian ethnic minority. The Armenians, however, became the subject of deepening distrust and were seen by many of the Young Turks as an obstacle on the path to national and nationalist modernisation. In 1915–1916, this led to a programme of mass deportation of the Armenians

– east, into the Syrian Desert – facilitated by Böttrich but contested and questioned by several other disapproving German officials stationed elsewhere around Turkey. The Baghdad Railway contributed directly to the genocide of the Armenians. 'Thus,' observes one careful historian, albeit in rather tortured wording, 'the Ottoman government introduced into modern history railway transport of civilian populations towards extermination.'[7]

But it was the Holocaust that saw the most deliberate, planned and scientific use of the railway as an instrument for extermination and evil distilled down to its very essence. Millions of Jews were transported to slave labour camps, and more especially the death camps, as just so many inanimate pieces (*Stücke*) in cattle trucks or locked boxcars (*Güterwagen*) aboard special trains (*Sonderzüge*). Deported without food or water and having access to only rudimentary ablutions facilities if at all, Jewish men, women and children experienced the terror of crammed mobile incarceration, and the humiliation of having to urinate and defecate in confined spaces, and were often forced into sharing the ultimate journey of horror in overcrowded conditions along with corpses and, occasionally, the insane.[8]

No WNLA train ever transported so much as a single Sul do Save migrant to a death camp, and there is an element of obscenity in placing the experiences of Mozambican miners on the WNLA trains on a continuum that might, in some way, imply a parallel with Armenians and Jews being transported by rail to sites of mass murder. In truth, the situations simply cannot be compared. Common means do not lead inevitably to common ends. That said, however, it is difficult to shake off entirely the uncomfortable feeling that there may well be some lingering truth to Hannah Arendt's provocative observation that there was something in the practices of colonialism that might have helped prepare the way mentally for the European barbarisms of the 20th century.[9]

The manner in which the WNLA trains were officially sanctioned and operated, first under British colonial and imperial administrations, later by English-speaking South Africans comfortable with increasing urban racial segregation, and then by Afrikaner nationalists in pursuit of grand apartheid, points to shared patterns of degraded and dehumanising thinking by whites that might, if intentionally extended into ethnic hatred and a civil war that could assume many forms, possibly have carried within it the potential for genocide. The fact that those controlling the migrant labour system were *not* intent on genocide did not, however, mean that the WNLA trains did not leave lasting scars on the consciousness of poorly educated rural Africans throughout the region in ways that went well beyond latter-day urban sophisticates' appreciation of the musical evocations of 'Shosholoza' and, of course, the haunting images in Hugh Masekela's 'Stimela'. In many rural minds, the ghosts of the dead and living dead are closely linked to the image of the train in an ultra-exploitative, modernising dispensation.

At Komatipoort, the Eastern Main Line used to be joined by a branch line that once carried WNLA recruits from Pietersburg in the far Northern Transvaal and then passed through the Acornhoek district of the Eastern Transvaal before debouching at the border complex. Acornhoek was thus not only integrated into the regional migrant-labour system by rail but also boasted large numbers of Shangaan residents, including many families who originally hailed from the Sul do Save and who, over time, became South Africans. Unsurprisingly, the WNLA trains, and more especially the 1949 disaster at Waterval Boven, were well remembered by many in the Acornhoek district. From at least the 1950s – and possibly even long before that – technology in the form of the steam locomotive and the train on the one hand, and notions of colonial subjugation and state power on the other, were interwoven into a set of fantastic sorcery and witchcraft beliefs that gained their clearest expression in Northern Sotho stories about *stimela sa baloi* – the witches' train.[10] In a way that can now be illustrated from many corners of the African continent, discourses around witchcraft cannot be simply dismissed as bizarre and frequently reflect some of the horrors of history – not excluding the precolonial history of slavery – in vivid terms.[11]

No one familiar with the dark history of the WNLA trains and the ferrying of mine labourers to Johannesburg, and the returning of the dead and the living dead to Ressano Garcia, can now read a remarkable account of the operations of the *stimela sa baloi* without seeing in it all the evil, including the potential for the wholesale destruction of human life, in a merciless migrant labour system serving primary industry. These witchcraft beliefs, which were still circulating freely through much of the Acornhoek district between 1978 and 1985 – some replete with tales about abductions and men simply disappearing in and around the Witbank collieries – confirm that, in the southern African countryside, both the past and the nightmare of the night trains are still very much with us. It seems fitting then that the final word on those trains should come from the mouths of those who either experienced them directly, or continue to be horrified by what they were told by those around them. So it is recorded in a comprehensive survey of late 20th-century witchcraft beliefs around Acornhoek that

Witches' trains resembled ordinary trains, but did not travel on rails. They were large, had many coaches, transported hundreds of passengers, and were staffed by personnel dressed in uniforms of the South African Railways. Their passengers, conductors and drivers were purported to be zombies. Witches' trains were hidden during the day, but were used at night to ferry zombies to their workplaces

193

... 'The zombies work in shifts just like mine workers'. While some are transported to work by the train, others are returned to their homes. Sometimes many witches collectively owned a single train, each using a particular coach for his or her zombies. Witches' trains abducted people who wandered about at night. Should they board, the conductor would ask them 'single or return'? Those who replied 'single' disappeared forever. They were killed, joined the zombies on the train and were forced to work for the witch. Those who replied 'return' were beaten and thrown from the train at a distant location.[12]

Such stories of exploitation and torture among subject peoples, suggests one analyst, can profitably be viewed as 'a colonial mirror that reflects back onto the colonists the barbarity of their own social relations'.[13] The story of the transport and exploitation of Mozambican migrant labour in the South African mining industry is, in good measure, a tale of a parallel universe, one deliberately concealed and, with the passing of time, one now in danger of becoming completely forgotten. It is time to look squarely in that colonial mirror and remember that the mirror does not lie.

The labourers drawn from the Sul do Save were recruited out of sight, in the seclusion of a far-off rural world through means that cannot be readily reconciled with the operation of a supposedly 'free labour market'. They were then transported – largely by night – in privately operated trains that denied their status as passengers and debouched at stations often obscured from the eyes of the South African public. Black workers, bound by contracts that were criminally enforceable, were medically examined and then moved, under escort, into company-policed housing – mine compounds – that to all intents and purposes were sealed off from almost all forms of social intercourse with outsiders and situated well beyond the scrutiny of all bar a few selected state officials. Within the confines of so repressive a social and working regime a significant number of workers retreated even further, into an alcohol- or drug-induced inner mental world.

A significant part of the workers' wages, held back in the form of gold bullion, was handed to a third party in Johannesburg by the Chamber of Mines, sold at a premium in the distant world markets of complicit northern-hemisphere financial agents and the difference pocketed by the Portuguese colonial administration back in Lourenço Marques. The remuneration of the migrant workers, too, was part of a largely hidden, manipulative and poorly understood system. For illiterate, poorly edu-cated Africans denied by law the right to organise workers at the point of production in a foreign country, it was one more part of a largely clandes-tine exploitative system supposedly devoted to a 'civilising mission' but one that they rightly saw as more akin to a form of slavery.

On completion of their contracts, the healthy among the workers – always far fewer in number than those who had left Mozambique – were entrained for the night journey back to the Sul do Save. At every station on the way back, and on the train itself, the miners were preyed upon by black and white South Africans intent on separating them by means, fair or foul, from that residual portion of their wages that they were carrying home in cash. When the night trains, denied the operational privileges and status of regular passenger trains in a system predicated on race, were involved in fatal accidents the names of some victims remained simply 'unknown'.

Those repatriates broken in body, and some in mind too – the industry's so-called rejects – did not qualify as 'patients' and were lifted into hospital coaches sporting the sign of the Red Cross but lacking entirely in doctors or fully qualified nurses. The insane, like those suffering from infectious diseases, were transported in coaches with barred windows. Concealment, camouflage and deception were integral to the functioning and longevity

*SAR hospital coach (top) and infectious disease coach (bottom). In the top photo,
note the barred window for 'mental cases' situated in the middle of the coach.*

of the night trains. Numbers of 'rejects' died on the train and corpses were
stripped of their wages before bodies were handed to over to the police for
post mortems conducted by district surgeons in small rail-side towns and
the lungs removed for medical research on the Rand.

So, where might all that leave the more thoughtful citizens of the most
advanced and diversified economy in Africa? Members of the new political

elite in a democratic South Africa have inherited a few limited, yet important, historical and political responsibilities that derive from a modern
infrastructure that was financed, in good measure, by the cheap labour
of southern Mozambican miners who, in numerical terms, outnumbered
that provided by any other grouping of colour in southern Africa. The
inhabitants of the Sul do Save, now one of the poorest regions on earth,
may have entered the modern world poor, but the mining revolution in
South Africa helped keep them and their descendants poor.

For those interested in truth and reconciliation in the greater southern
African region, the moment may just have arrived to think about the role
that predatory capitalism, the night trains and an always compliant South
African state played in a socio-economic history that is as integrated as
it is troubled. All South Africans, but more especially those who owned
and own the coal- and gold-mining industries, need to acknowledge that
much of the country's past prosperity, wealth and relatively advanced
infrastructure were built on the backs of black labour pushed and pulled
out of colonial Mozambique.

Those in South Africa who eschew xenophobia, those who see beyond
the accepted borders of a nation-state won in a liberation struggle sup-
ported by Mozambicans and built in good part by workers from the Sul
do Save, those who preach continental and regional solidarity, along with
those who trumpet the virtues of ubuntu, may wish to pause and think
again about the unique contribution that the men, women and children
of southern Mozambique made to the evolution of modern South Africa,
and how the night trains helped maintain the Sul do Save as an economic
backwater of globally scandalous proportions. Those South Africans
rightly interested in reparations being made for past injustices – economic
and social – may need to adopt a slightly broader perspective, one more
in keeping with the spirit evoked while singing 'Shosholoza' or 'Stimela'.

# Afterword

*The real voyage of discovery consists not in seeking new lands
but seeing with new eyes.*

— PROUST, *REMEMBRANCE OF THINGS PAST*

When I was about ten years old, back in the mid-1950s, I spent many an afternoon, and sometimes an entire Saturday, at the remote Springvale stores, opposite a compound and mine shaft, not far from modern-day Stilfontein in what then was still the Western Transvaal. Three 'native' stores stood there, fringed by a smooth concrete-floored verandah that, partly shielded by a protruding corrugated-iron roof, provided some relief from a Sahara-like sun intent on murder. Behind the stores, perched two feet above the dusty and rutted sand road that ran in front of the place, the acacia bush had been cleared for about 50 metres, but, beyond that, the scrub remained triumphant. Black miners who patronised the only stores within walking distance of the compound often relieved themselves there, behind a thorny green screen. Beyond that, beneath one or two thorn trees tall enough to pretend that they were not really acacias, scatterings of miners would sit and drink home-brew.

Springvale was a lonely place for a lad. I never relished going there but my father did because a close friend owned all three shops, and they spent many hours seated in the back of the main store, chatting, eating and drinking. The three stores – a butcher shop, a 'concession store' and a 'native eating house' – were interlinked, and one could walk from one end of the complex to the other without ever having to enter through the front doors of the stores.

The hub of the complex, the concession store – so named because it stood on leased mine property and was thus a 'concession' to the owner by the company that owned everything within easy sight and just about as far as the eye could see – was the jewel in the crown, and an inquisitive boy's delight. Towards the front it contained every cheap, bright and often shoddily made item in the known world – from stacks of tins of sweetened condensed milk and pilchards and small packets of tea and sugar, to hand mirrors, razors and pocket- and hunting-knives, as well as mouth organs in several sizes, tin whistles and a few piano accordions. Beyond the small, locked glass counters holding most of the boy-treasures, there were shelves struggling to contain all the shirts, trousers, ties, cravats and scarves, caps and hats that you could imagine. Affordability and gaudiness were locked in a tight embrace. But the most striking section, by far, was at the far end of the store where, seen through a child's eyes, there seemed to be acres of floor space and shelving that reached truly monstrous proportions. Stacked high were dozens, scores – perhaps a hundred – tin trunks of various sizes, all bearing handles that looked inadequate to carry the weighty contents and metal staples meant for locks so weak that they could not possibly have been related to the venerable Yale family. But it was the floor-to-ceiling shelves, neatly stacked with woollen blankets, perfectly aligned – all exactly the same size – in brilliant colours and exotic designs that imprinted themselves on the minds of nearly all the visitors.

It did not take long for a boy bored and craving company – any company – to notice that the miners did not emerge from the guarded compound gates to cross the dirt track leading to the store in anything like a steady stream. True, there were always one or two stragglers wearing plastic wristlets bearing numbers hovering about on the verandah outside the shops – and more especially the eating house – but, for most of the day, there was no easily discernible pattern as to what was happening: no early-morning customers, no lunch-time shoppers or late-afternoon rush before the stores closed around sunset.

It seemed like a place where there was no notion of time that accorded with convention. My father had a wristwatch that, when home, he paid attention to, but not one of the 'natives' possessed a wristwatch – not even a rounded old pocket watch – or, if they did, they never bothered to consult them. At Springvale, time was suspended and the sun – for all its menace – did not manage to impart much meaning to men's movements. It was bewildering for a boy intent on establishing where the balance between time-elapsed and time-still-to-go lay before getting back to the comforts of the family home.

At first, these mildly disorientating Springvale experiences – occasioned by the absence of familiar pointers encountered at home, such as day giving way to night, holidays and weekdays, church or sport, play or school, and meal or tea-times – were exacerbated by an unpredictable, exciting, occurrence around the mine stores. Every now and then, the steady trickle of miners out of the compound gates, usually in ones or twos – for reasons that were unfathomable – suddenly assumed torrential proportions. Whenever that happened, the otherwise stark and isolated stores that seemed to rise about the surrounding veld like bared rocks in a shallow seaside gully were quickly inundated by the rising tide of African men. Hundreds of miners, having just finished showering, swarmed around the butcher shop, buying the cheapest cuts of beef and offal available or, if too exhausted to think about cooking their own meat, wriggled into the rows of long wooden benches behind metal-topped tables in the eating house to consume large piles of stodgy maize porridge topped by browned stew always dripping fatty gravy regardless of the season. It was only after the inner men's cravings for food had been briefly stemmed that tired bodies were allowed to drift off to the concession stores to buy 'luxuries'.

On Saturdays at Springvale, I longed for that high tide of endlessly hungry miners, which, I eventually worked out, usually arrived in mid-morning or midafternoon. Then the stores and the sullen surrounding scrub erupted into life and everything inside the butcher shop and eating house, as well as behind them, where meat was roasted on open fires, became a bit frantic. Inside the concession store, however, the atmosphere was markedly more measured and subdued. Once the beast-in-the-belly had been beaten off for a few hours with the help of some hard-earned cash spent on *pap*, the mind regained control of body and soul and any additional money was dispensed only after the most careful thought and much deliberation. A man could spend half an hour looking, thinking and touching before buying a mouth organ worth a half-crown. Sometimes, when pounds rather than shillings were called for, the worker would come and inspect the blanket of his dreams on two, even three successive weekends before taking the plunge and buying a supposed luxury.

For some reason or other, or so it seemed to me, the midmorning miners were always more famished and physically spent than were the midafternoon men. The butcher shop and eating house usually had more customers before midday than after. I got to understand how the waves of exhaustion were linked to the incoming tide only after the friendly black clerk at the concession store unravelled the mysteries of shift work for me. I found the idea that men were expected to work underground from 10 pm right through to 6 am the following morning horrifying, and a grudging respect for the mere physical prowess of black miners working for very low wages mutated into an element of admiration for their undoubted mental fortitude.

Although still almost totally myopic, I slowly, oh so very, very, slowly, began to see the miners in a new and slightly different light, but since few others around me seemed to pick up on what I was beginning to 'see' – although not 'conceptualise' – as an utterly merciless system of black labour exploitation, I tended to keep my views to myself. It would be pleasing now if I could say that I then began to befriend or speak to the Springvale miners. I did not. How could I? They were all in their twenties, grown men hailing from exotic places that I knew only vaguely from collecting stamps – Basutoland, Portuguese East Africa, Swaziland and so on – who chatted in tongues so foreign that not even the concession-store attendants could communicate with them in anything other than *fanakalo*, that stunted master-servant *lingua franca* of the mining industry.

I never got to 'know' a single black miner at Springvale, but I did get to recognise a few of them, and they me. Once they cleared the sawdust-covered floor of the butcher shop, or emerged from behind the greasy surfaces of the eating house tables, I latched onto them outside the entrance and followed them into the concession store, dogging their every step, observing their choice of items and their deliberate movements with endless curiosity.

The black miners, their limbs often gleaming beneath a thin layer of Vaseline, all came in the standard shape and size demanded by the industry – lean and strong – and, for the most part, they looked to me to be as tall as did every other grown-up. Some arrived at the stores on their own, looking like loners, and, by the same token, were seldom promising targets for a small human magnet intent on fixing itself onto anyone who appeared remotely social. There were almost always one or two others whom I had been cautioned about – but seldom more – who came into the store as a couple, openly affectionate with one another and, not infrequently, holding hands. Most promising of all, however, were small clusters – the best only three or four strong – of cheerfully relaxed tribesmen who, to me, seemed to exude friendliness, tolerance and patience.

I would stand around the fire behind the concession store with them, clutching a curved glass bottle of Coke, watching them slurping soured milk from machine-assembled elongated cardboard containers and waiting for their blackened, smoke-stained meat to finish roasting. I chewed bubble gum that came in wrappers containing quiz questions, which intrigued me, while they deftly fingered-in large mouthfuls of porridge, by then cold, between slivers of hot meat. But the best, shared moments between what might superficially pass as 'friends' always came on cold winter mornings, when the wind was icy and the last of the frost crystals clung to the long-grass beard of the scrub.

Behind the butcher shop stood two large, well-scoured 44-gallon drums, and beneath them a small fire was fed pieces of gnarled acacia from the time that the sun appeared until it eventually tired of watching and disappeared in a rosy haze over the desert to the west. In between, large pitted aluminium tubs – *skottels* – brimming with chunks of snow-white animal fat, containing tiny nuggets of sinewy meat that had escaped the probing of the butcher's knife, appeared from the back door of the butchery at regular intervals. The fat was eased into the drums and bubbled away for hours until it was all rendered as lard.

The fire provided warmth, the bubbling mixture an indescribable odour that some might have found appealing, and the chattering black men circling the drum fine, good-humoured company. It was part of a simple, rewarding waiting game. Every now and then the butcher's assistant would plunge a long-handled ladle into the boiling fat, haul out a half-dozen chunks of sizzling crackling, and toss them into an empty *skottel*. The key to success was twofold: first, in promoting yourself to the inner circle, so that you would not have to reach out over the arms of others when you stretched out to grab the hissing crackling, and, second, having fingers tough enough to hold on to the prize, since anything dropped would be scooped up by those in the second tier.

The delight of chewing, eating and licking a crunch of hot crackling from my lips in the outdoors on the coldest of winter mornings, in the company of giants who formed a protective circle around me, stays with me as an experience as if it were yesterday. So, too, is the fact that I was much better than any of those friendly miners at getting the finest of the crackling prizes because I always managed to wriggle into the inner circle ahead of any of them and I clearly had tougher fingers than any old miner. Or so it seemed to me way back then, and I have no desire to change that perception now or ever.

Springvale is now six decades and a few years behind me, yet I know that even back then I had two big questions about black miners that have haunted me all my life. The first of these was who precisely those 'foreign natives' were, where exactly they came from, how they got themselves to the mines and returned home, and why they would choose to endure such a miserable life away from their families and loved ones in exchange for blankets? There was something inexplicable, something illogical, at work in a system that I could 'see' but could never fully understand. The second question was about what happened inside the off-limits mine compound and how the men there lived in a space that I was never allowed to enter and that was guarded by compound police. Why was that? What was happening inside that sealed-off square?

For some years after that, black miners, mine compounds and labour systems went out of my consciousness as I struggled to get through high school. When my father asked the principal whether I might gain admission to a university, the principal told him that he could not, in all honesty, recommend that such limited funds as he had be wasted on an attempt to gain a tertiary education for his very average son. My father had once been a policeman, and my mother, who had never completed her schooling, worked in a shoe factory when she left school at age 15. I would have been the first from either side of my family to attend a university.

But then, at the youthful age of 16, I did indeed go off to university, on interest-accruing funds borrowed from the state. Then, having barely scraped through the first-year examinations and having just turned 17, I returned home only to be told that, if I wished to continue my studies, I would have to spend the next three months working on the mines in order to help ease the financial burden.

Barely half a decade had passed since I had been nibbling on crackling at Springvale. I was an awkward, overgrown, hopelessly immature schoolboy, and yet there I suddenly was – a would-be young graduate working night shifts as an onsetter underground in one of the deepest gold mines in the world, alongside some of the most brutish and toughest white men in the country. My dislike of most of the white miners, the nature of the work that I had to do – hauling men and equipment from the surface underground in a small 'cage' – and a hatred of the way the industry was racially ordered and run, was matched only by my cowardice, my fear of danger and my silence about injustices.

Once again, maybe for slightly different reasons that time round, I failed to befriend so much as a single black miner. I tried to be fair and reasonable when dealing with them, but I never spoke to any them in any other way than was expected of a young white 'man' briefly 'in charge' of black men at work. I did, however, gain a few more insights as to how the industry was managed and organised, even though I was still intellectually ill-equipped to deal with the two questions that had haunted me since Springvale days: what was the purpose and function of the compound system, and how and why did the black miners continually shuttle back and forth between their homes and the mines?

About ten years after those later experiences on the Free State gold mines, in the late 1960s, there followed a few more largely unsuccessful attempts at university to discover some – any – sort any intellectual competence and coherence in my efforts to understand more about the world. More spells of much-resented work underground ensued as I battled to pay back the state loan that dragged me down into the mines. Then, belatedly, I stumbled into a love affair with history and everything changed. I say 'stumbled', because I had never studied history formally at high school, or by enrolling for a course at tertiary level. It was history that gave me 'new eyes' and new ways of making sense of my own life experiences, and those of others.

When I eventually got round to preparing myself for a career as a professional historian, I struggled to find a research project for a doctoral study. In those days, in the early 1970s, the regulations required an original argument that had to be presented in sufficiently convincing terms, and in a format that might lend itself to immediate publication as a book by a commercial publisher. After floundering for some weeks, one day I recalled the visits to Springvale and decided that I would try to answer the first of the two big questions that had plagued me as a boy. It culminated in a study of the mine compound system and, to my delight, was quickly taken up and published as an academic book.

My examination of the mine compound system was deliberately *not* set in South Africa, but rather in colonial Zimbabwe – in Southern Rhodesia – or, more accurately, within the broader context of the regional economic system. I have always disapproved – and still do – of historical studies that uncritically assume the nation-state and its boundaries as an unproblematic unit of analysis. To me, the state, in all its singular glory and power, constitutes an inappropriate and unhelpful analytical entity when trying to understand more fully interlocking colonial systems that are characterised by significant flows of cheap, transnational migratory labour underwritten by people who are racially, politically and socially marginalised when not overtly oppressed. In Africa, of all places, arbitrarily drawn colonial borders that make little sense in either ethnic or geographical terms seem to offer an exceptionally weak platform on which to try and construct an understanding of political economies and new societies.

Nation-state analyses need not, but in practice frequently do, feed into the narrative of dominant nationalist elites, who in my experience often subvert historical writings for purposes that are primarily political rather than to advance understanding or clarify our thinking. For much the same reason – a personal aversion to ethnic nationalists and to a nationalist propensity to selectively appropriate historical writings – the book that I wrote next tried to circumvent some of these problems. I tried to explore the history of economic and social groupings who were locked into a mining metropolis but *not* directly dependent on the mining industry for their livelihoods, and did so in ways that consistently underscored the importance of ethnic, religious and national diversity, as well as multiculturalism, multiracialism and the dynamics of class.

But that, too, was in the 1970s, and after that I paid scant attention to the important role that cheap, transnational, African migratory labour played in the historical development of South Africa's stalled, if not failed, industrial revolution. I was far more interested in the dynamics of two other interlinked processes that, for me, explain a lot more about the nature of developed and underdeveloped economies and societies in the late 19th and much of the 20th centuries, that is, the phenomena of crime-as-politics and politics-as-crime.

The second question, of how and why black miners from all round the southern African region came to work on the gold mines and then returned to their homes remained unopened for several decades, like a packet of seeds long forgotten in a drawer somewhere. Recently, however, I embarked on the research for a final, large project before I settle into my dotage: trying to establish how, in functional and historical terms, Anglophone, urban, industrial and Protestant South Africa has related to Lusophone, rural, commercial and Catholic Mozambique during the 20th century. No such study would be complete without paying great attention to the historical importance of labour migration from Mozambique to South Africa during the high point of a regional industrial revolution that was as fractured and uneven as it was unsuccessful.

This book is a reasonably self-contained outgrowth of the larger work that will, hopefully, follow in due course. It examines just one very limited aspect of the migratory experiences of African men drawn from just one part of Mozambique, the Sul do Save, and that is the return train journey between the border town of Ressano Garcia and the metropolis of Johannesburg. It may or may not fire the imagination of readers who are interested in the evolution of the wider region. It would be nice if it did, but in my experience we are living through times when the ships of history that rely on the winds of genuine curiosity about the past to drive them – as opposed to the carts of 'heritage' dragged by plodding nationalist horses – no longer work for those sailing into bookshops. More importantly, remember that this book is part of the answer to a few bigger questions that puzzled a ten-year-old more than 60 years ago.

# Acknowledgements

As with previous works, I have been encouraged by many exceptionally generous and thoughtful colleagues, friends and close family members who have given unstintingly of their time and provided me with helpful and thoughtful criticism. If there are any merits to what remains it is, in large measure, thanks to them. Any remaining errors in evidence or interpretation are all my own work. Should I have inadvertently forgotten to name any one who has contributed directly or indirectly to the completion of this book, then I can only plead guilty to all charges and beg for forgiveness.

The trick of resorting to alphabetical order to arrange the names of the few who have done so much to help me with serious questions and their expertise, and the many who have assisted with practical professional help in a host of other ways, is unfair. But fairness is always more of a worthy aspiration than an accomplishment and this list will, alas, be no exception. The brief list of acknowledgements and gratitude expressed below is, I fear, in inverse proportion to the benefit that I have received from all those named. It is my hope that they think that the exchange has been worthwhile:

E Ashton, J Ball, C Bailie, J Boraine, B Bozzoli, D Ballantyne, R Crewe, C de la Rey, RM Godsell, K Harris, HB Heymans, D James, P La Hausse,

I Niehaus, AH Jeeves, R Kaplan, A LeMaitre, M Mubai, T Mugabe, J Ogude, P Pearson, J Penvenne, GH Pirie, IR Phimister, C Prenter, M Shain, JD Sinclair, B Strydom, M Vaughan and a clutch of four van Onselen captives bearing the intials C, G, J and M, along with the genial and tolerant 'Friends of the Rainbow Room', whose names for legal and other reasons can never be disclosed publicly.

For the illustrations that appear in this book – maps, photographs and technical drawings – I am more especially indebted to the following people who not only were generous in sharing their vast railways expertise with me, but gave me and the publishers permission to use material drawn from specialist private collections assembled over many decades: HB Heymans, B Martins, Y Meyer, RCJ Pretorius, L Pivic, P Stickler and J and J Wepener. The technical drawings are copies of originals that were once preserved in the drawing office of the South African Railways & Harbours building in Pretoria. The inspiration for the fine illustrated dust jacket derives from a marvellously evocative painting by Wendy Malan. I would also like to thank B Masekela and P Twala and members of the Hugh Masekela Heritage Trust for permission to reproduce the lyrics of 'Stimela'.

\* \* \*

Some time back I made what, back then, seemed to be the difficult decision to leave an institution where I had been employed for many years but which, in its opportunistic haste to reach an accommodation with the incoming political elite as the old nationalist order gave way to the new, was, by then, impatient to obliterate any remaining signs of such academic work as I and several friends and professionals had undertaken there over two preceding decades. It shut down an institute that had been many years in the making and produced many fine scholarly works. It took a while for me to know whether or not I had taken the right fork in the road that opened up before me at that time. Each passing year now confirms that I did.

The nearly twenty years I have spent in the Centre for the Advancement of Scholarship (CAS) at the University of Pretoria researching and writing history have been the most contented, fulfilling and productive in a career that, in terms of formal professional association with universities, is now, alas, gradually drawing to a close. I could not have asked for more encouragement, support for, or recognition of my attempts to write history than I have received from the past and present directors of the CAS and several vice chancellors of the University of Pretoria, or from my long-suffering publishers, Jonathan Ball.

All works of history are products of their times and eventually need to be rewritten as new insights, questions and sources emerge, but all historians have to aspire to writing about the past in ways that cannot readily be appropriated by the ruling classes of the day who wish to make knowledge about the past the servant of a present that they alone wish to define and inflict on the citizenry. Good history raises questions and should serve as a health warning about all past and present government-approved histories and enjoy a bookshelf life that outlasts several generations of academic fashionistas. Let us see how this one ages.

*Charles van Onselen*
*Parkview*
*June 2019*

# Notes

## INTRODUCTION

1 See AP Cartwright, *The Gold Miners* (Cape Town, 1961), pp 216–222 [hereafter Cartwright, *The Gold Miners*].

2 Cartwright, *The Gold Miners*, p 216.

3 The questions around the 1949 Waterval Boven train disaster, along with contemporary 'heritage' sites there, require political rather than scholarly answers. In due course I will address what has happened there in a polemic rather than in this work, which explores other wide-ranging issues relating to the labour trains.

## CHAPTER ONE

1 Slightly modified from a fragment first presented to American readers by HA Junod who, in *The Life of a South African Tribe* (New York, 1962), p 197, wrote about 'the braves' rather than warriors.

2 Hugh Masekela, son of a sanitary inspector father and a social worker mother in KwaGuqa, who went on to spend some of his youth with an uncle based at Rose Deep Mine, in Germiston, was ideally placed to observe train and miners alike; C van Onselen, interview with Hugh Masekela, 27 September 2017.

## CHAPTER TWO

1 See WA de Klerk, *The Puritans in Africa: The Story of Afrikanerdom* (London, 1975), pp 6–61.

2 See A Porter, *Victorian Shipping, Business and Imperial Policy: Donald Currie, the Castle Line and Southern Africa* (Woodbridge, 1986), pp 164–166.

3 See SE Katzenellenbogen, *South Africa and Southern Mozambique: Labour, Railways and Trade in the Making of a Relationship* (Manchester, 1982), pp 20–21 [hereafter Katzenellenbogen, *South Africa and Southern Mozambique*].

4 See CT Gordon, *The Growth of Boer Opposition to Kruger, 1890–1895* (London, 1970), pp 58–59 [hereafter Gordon, *The Growth of Boer Opposition*], and Katzenellenbogen, *South Africa and Southern Mozambique*, pp 16–17.

5 For some of the social history associated with the construction of the Eastern Main Line, see C van Onselen, *Showdown at the Red Lion: The Life and Times of Jack McLoughlin, 1859–1910* (Johannesburg, 2015), pp 152–158.

6 On the NZASM's early financial and other problems see, for example, Gordon, *The Growth of Boer Opposition*, pp 58–61.

7 The global timber trade and links to the Transvaal coal and gold mining industries in the late 19th and 20th centuries remain surprisingly neglected topics among economic historians; see also C van Onselen, *The Cowboy Capitalist: John Hays Hammond, the American West & the Jameson Raid* (Johannesburg, 2017), p 43.

8 See GH Pirie, 'Railways and Labour Migration to the Rand Mines: Constraints and Significance', *Journal of Southern African Studies*, Vol 19, No 4 (December 1993), p 718.

9 See especially J Haarhoff, 'Conquering the Escarpment: Railway Engineering in the Elands River Valley', *Civil Engineering* (May 2017), pp 20–30.

10 See, for example, 'Dining Car Arrangements', *Rand Daily Mail*, 11 December 1922.

11 See Republic of South Africa (RSA), University of Johannesburg Library (UJL), TEBA Collection, WNLA File 33A, Secretary, Transvaal Chamber of Mines to General Manager, South African Railways & Harbours, 29 April 1936.

# CHAPTER THREE

1 C van Onselen, *New Babylon, New Nineveh: Everyday Life on the Witwatersrand, 1886–1914* (Johannesburg, 2001), pp 1–2.

2 See SH Frankel, *Investment and the Return to Equity Capital in the South African Gold Mining Industry, 1887–1965* (Cambridge, MA, 1967), pp 7–8 and 127. I am indebted to Ian Phimister for providing me with an idiot's guide to the sometimes opaque commentary that accompanies Frankel's extraordinary calculations.

3 See AH Jeeves, *Migrant Labour in South Africa's Mining Economy; The Struggle for the Gold Mines' Labour Supply, 1890–1985* (Johannesburg, 1985), p 121 [hereafter Jeeves, *Migrant Labour*].

4 This paragraph is based on data drawn from F Wilson, *Labour in the South African Gold Mines, 1911–1969* (Cambridge, 1972), pp 9–10, 40 and 46 [hereafter Wilson, *South African Gold Mines*]. See also J Crush, A Jeeves and D Yudelman, *South Africa's Labor Empire: A History of Black Migrancy to the Gold Mines* (Cape Town, 1991), p 3 [hereafter Crush et al, *Labor Empire*].

5  R First, *Black Gold: The Mozambican Miner, Proletarian and Peasant* (New York, 1983), pp 25–26.

6  See A Isaacman and B Isaacman, *Mozambique: From Colonialism to Revolution, 1900–1981* (Aldershot, 1983), p 33 [hereafter Isaacman, *Mozambique*].

7  L Vail and L White, *Capitalism and Colonialism in Mozambique: A Study of Quelimane District* (London, 1980), p 206 [hereafter Vail and White, *Capitalism and Colonialism*].

8  W Gemmill, as quoted in Vail and White, *Capitalism and Colonialism*, p 206.

9  Wilson, *South African Gold Mines*, p 70; D Yudelman and A Jeeves, 'New Labour Frontiers for Old: Black Migrants to the South African Gold Mines, 1920–1985', *Journal of Southern African Studies*, Vol 13, No 1 (October 1986), pp 102–104; and J Penvenne, *African Workers and Colonial Racism: Mozambican Strategies and Struggles in Lourenço Marques, 1877–1962* (London, 1985), p 96 [hereafter Penvenne, *African Workers and Colonial Racism*].

10  See 'Native Labour Association', *Rand Daily Mail*, 4 April 1913. Propaganda about the alleged 'popularity' of mine labour – an allegation that slowly acquired more traction on the truth as the subsistence economy of rural Mozambique continued to decline – persisted well into the interwar years. In 1919, for example, a spokesman for the industry said that Africans were 'attracted to the mines simply by reason of the excellent conditions of employment' – 'Not "Wiped out by Disease"', *Rand Daily Mail*, 1 April 1929.

11  On rational choices made by African miners within the regional economy, including southern Mozambican Shangaans, see, for example, C van Onselen, *Chibaro: African Mine Labour in Southern Rhodesia, 1900–1933* (London, 1976), pp 86–91 [hereafter van Onselen, *Chibaro*]. And, for choices made in several different settings over a much longer period, see P Harries, *Work, Culture, Identity and Migrant Labourers in Mozambique and South Africa, c 1860–1910* (Johannesburg, 1994), pp 19–80 [hereafter Harries, *Work, Culture, Identity*].

12  There is agreement among historians of almost all persuasions about how a coercive and forced labour regime in Mozambique, and *shibalo* in particular, drove Thonga or 'Shangaan' labour to the Rand mines. See, for example, Harries, *Work, Culture, Identity*, p 147; or SE Katzenellenbogen, *South Africa and Southern Mozambique: Labour, Railways and Trade in the Making of a Relationship* (Manchester, 1982), p 60 [hereafter Katzenellenbogen, *South Africa and Southern Mozambique*]; and M Harris 'Labour Emigration among the Moçambique Thonga: Cultural and Political Factors', *Africa*, Vol 29, No 1 (January 1959), p 60 [hereafter Harris, 'Labour Emigration'].

13  On the slave trade in Mozambique, see especially J Duffy, *Portuguese Africa* (Cambridge, MA, 1959), pp 132–152 [hereafter Duffy, *Portuguese Africa*]; and G Clarence-Smith, *The Third Portuguese Empire, 1825–1975* (Manchester, 1985), pp 34–35 and 76 [hereafter Clarence-Smith, *The Third Portuguese Empire*]; and, on indentured labour in the 1850s, Vail and White, *Capitalism and Colonialism*, pp 30–31.

14  United States of America, California, Stanford University Library, Despatches from

United States Consuls in Mozambique, 1854–1898, Roll 1, Vol 1, S Hollis to The Assistant Secretary of State, Washington, DC, 23 November 1894.

15 See Duffy, *Portuguese Africa*, p 149; Vail and White, *Capitalism and Colonialism*, p 24; and M Newitt, *A Short History of Mozambique* (Johannesburg, 2018), pp 68–69 [hereafter Newitt, *Short History*].

16 For chained labourers from Inhambane being sent to the Kimberley diamond mines in the early 1880s, see Harries, *Work, Culture, Identity*, p 54.

17 Harries, *Work, Culture, Identity*, p 127.

18 Van Onselen, *Chibaro*, p 99 and Katzenellenbogen, *South Africa and Southern Mozambique*, p 58. Nor were Africans the only ones to see this type of linkage at the time. See, for example, HW Nevinson, 'The New Slave Trade', *Harper's Monthly Magazine*, Vol 111, No 663 (August 1905), pp 349–350.

19 See Harries, *Work, Culture, Identity*, p 142.

20 Seen from an early Calvinist perspective, such enforced self-discipline could be seen as a 'sign of freedom and agency rather than subjugation or repression', but by the 20th century such visions had long since been eclipsed; see PS Gorski, *The Disciplinary Revolution: Calvinism and the Rise of the State in Early Modern Europe* (London, 2003), p 166.

21 See E Allina, '"Fallacious Mirrors": Colonial Anxiety and Images of African Labor in Mozambique, c 1929', *History in Africa*, Vol 24 (1997), p 12; Duffy, *Portuguese Africa*, pp 150–155; Isaacman, *Mozambique*, p 34; and Newitt, *Short History*, pp 113–114 and 130.

22 Duffy, *Portuguese Africa*, p 156.

23 Clarence-Smith, *The Third Portuguese Empire*, p 12.

24 See Isaacman, *Mozambique*, pp 39–40; and Penvenne, *African Workers and Colonial Racism*, p 108.

25 Clarence-Smith, *The Third Portuguese Empire*, pp 181–182; and Newitt, *Short History*, pp 129–131.

26 HW Haynes, 'How the Industry Gets its Workers', *Rand Daily Mail*, 4 July 1933.

27 See, among others, Penvenne, *African Workers and Colonial Racism*, pp 17 and 25; O Roesch, 'Migrant Labour and Forced Labour Production in Southern Mozambique: The Colonial Peasantry of the Lower Limpopo Valley', *Journal of Southern African Studies*, Vol 17, No 2 (June 1991), pp 230–270; and Vail and White, *Capitalism and Colonialism*, pp 244 and 383.

28 See Clarence-Smith, *The Third Portuguese Empire*, p 109; Harris, 'Labour Emigration', p 60; Harries, *Work, Culture, Identity*, p 170; Jeeves, *Migrant Labour*, p 190; Katzenellenbogen, *South Africa and Southern Mozambique*, pp 40–41 and 102; Penvenne, *African Workers and Colonial Racism*, p 24; and Vail and White, *Capitalism and Colonialism*, pp 135–136, 207 and 249. Katzenellenbogen notes that this coupling of forced labour to work on the mines could at times be as crude as it was direct. Any black man having been recruited for the mines and deserting on his way to the Witwatersrand

automatically became liable for 90 days' *shibalo* labour.

29  Duffy, *Portuguese Africa*, p 170. See also Newitt, *Short History*, pp 105–106
30  Newitt, *Short History*, p 106.
31  Some analysts see the period of extreme physical coercion giving way to labour 'flowing freely' by as early as 1910; see Crush et al, *Labor Empire*, p 5.
32  These realities complicated and slowed down bilateral relations and negotiations. See, for example, PG Eidelberg, 'The Breakdown of the 1922 Lourenço Marques Port and Railway Negotiations', *The South African Historical Journal*, Vol 8, No 1 (2000), pp 104–118. For the entrenchment of South African dominance in the 1920s, see also Vail and White, *Capitalism and Colonialism*, pp 210–214 and 235.
33  See Katzenellenbogen, *South Africa and Southern Mozambique*, p 13; CF Spence, *The Portuguese Colony of Moçambique* (Cape Town, 1951), pp 85–87; and GE Vaughan, *Portuguese East Africa* (London, 1952), p 10.
34  Katzenellenbogen, *South Africa and Southern Mozambique*, pp 152–153.
35  See Jeeves, *Migrant Labour*, p 197; and Newitt, *Short History*, p 99.
36  See Jeeves, *Migrant Labour*, Table 6, p 191; Katzenellenbogen, *South Africa and Southern Mozambique*, p 66; and AK Smith, 'The Idea of Mozambique and its Enemies, c 1890–1930', *Journal of Southern African Studies*, Vol 17, No 3 (September 1991), p 510.
37  South African National Archives (SANA), Pretoria, NTS, Vol 201, 1911, Witwatersrand Native Labour Association Limited, 'Report of the WNLA Enquiry Commission' (Confidential), 1911, p 4.
38  Jeeves, *Migrant Labour*, p 27.
39  See, for example, 'A Slavery Lie' and 'Terrorism by Sypias', *Rand Daily Mail*, 2 February 1910 and 2 May 1913.
40  See Harries, *Work, Culture, Identity*, pp 174, 181; and Jeeves, *Migrant Labour*, p 93.
41  See GH Pirie, 'Railways and Labour Migration to the Rand Mines: Constraints and Significance', *Journal of Southern African Studies*, Vol 19, No 4 (December 1993), p 714.

## CHAPTER FOUR

1  Republic of South Africa (RSA), University of Johannesburg Library (UJL), TEBA Collection, File 33A/3, District Manager, WNLA, Lourenço Marques to General Manager, WNLA, Johannesburg, 1 October 1927.
2  See South African Railways & Harbours, annual *Reports of the General Manager*, covering the period 1910–1960.
3  For examples of the uses of these terms see RSA, UJL, TEBA Collection, File 33A, BL Thurber to Compound Manager, WNLA, Johannesburg, 1 June 1936 ('waybill'); File 37/1, WW Hoy, Acting General Manager, SAR to F Perry, WNLA, Johannesburg, 7 July 1910 ('imported'); and File 33A/7, Secretary, Transvaal Chamber of Mines to Acting General

Manager, SAR, 29 May 1936 ('human freight').

4  See RSA, UJL, TEBA Collection, File 37/1, Secretary, WNLA to General Manager, SAR, 5 January 1912 ('raw savages'); File 37/1, Secretary, WNLA to Acting General Manager, SAR, 29 June 1910 ('raw natives'); and File 33A/3, General Manager, WNLA to General Manager, SAR, 15 May 1959 ('primitive natives').

5  As quoted in AH Jeeves, *Migrant Labour in South Africa's Mining Economy* (Johannesburg, 1985), p 214.

6  See RSA, UJL, TEBA Collection, File 55/2 AS 69, Herbert C Gain (of Moorgate Hall, Finsbury, London EC) to Assistant Secretary, WNLA, Johannesburg, 30 May 1917 ('biscuits'); File 33A/5, General Manager, WNLA to General Manager, SAR, 12 April 1958 ('enable feeding'); and File 33A/3, Compound Manager, WNLA to General Manager, WNLA, 16 December 1926 ('stray natives').

7  P Harries, *Work, Culture, Identity and Migrant Labourers in Mozambique and South Africa, c 1860–1910* (Johannesburg, 1994), p 179, who writes – somewhat misleadingly – of 'sheep trucks and closed luggage vans'.

8  For the reluctant provision of drinking-water facilities, for example, see GH Pirie, 'Brutish Bombelas: Trains for Migrant Gold Miners in South Africa, c 1900–1925', *The Journal of Transport History*, No 18 (1997), pp 33–35 [hereafter Pirie, 'Brutish Bombelas'].

9  Republic of South Africa (RSA), South African National Archives (SANA), Pretoria, TAB Vol 183, File 933/14, Secretary, WNLA to Acting General Manager, SAR, 26 July 1911.

10  For examples see RSA, UJL, TEBA Collection, File 33A/3, Compound Manager to General Manager, WNLA, 4 April 1924, and Secretary, Transvaal Chamber of Mines to General Manager, SAR, 27 August 1928.

11  Pirie, 'Railways and Labour', p 729.

12  RSA, UJL, TEBA Collection, File 33A/4, General Manager, WNLA, Johannesburg to Manager, WNLA, Lourenço Marques, 14 January 1963 ('labour train'); File 33A, Secretary, Transvaal Chamber of Mines to General Manager, SAR, 29 April 1936 ('kaffir mail'); and File 33/A, Manager and Secretary, WNLA, Johannesburg to General Manager, SAR, 3 July 1923.

13  RSA, UJL, TEBA Collection, File 33A, R Thomson to Compound Manager, WNLA, Johannesburg, 9 June 1941.

14  See, for example, RSA, UJL, TEBA Collection, Manager and Secretary, WNLA, Johannesburg to General Manager, SAR, 29 October 1920.

15  RSA, UJL, TEBA Collection, File 33A/5, A Nosworthy to Compound Manager, WNLA, 21 July 1955.

16  Objections to the erection of the WNLA compound at Alto Mahé, in Lourenço Marques, in 1912 – strongly resisted by municipal authorities and residents alike – is a case in point; see RJ Williams, 'Creating a Healthy Colonial State in Mozambique, 1885–1915', unpublished PhD thesis, Department of History, University of Chicago, 2013, pp 238–239.

17  RSA, UJL, TEBA Collection, File 33A/3, BL Thurber to Compound Manager, WNLA, Johannesburg.

18  RSA, UJL, TEBA Collection, File 33A/3, System Manager, SAR to WNLA, Johannesburg, 27 March 1939.

19  RSA, UJL, TEBA Collection File 33A/3, JA Gemmill to the General Manager, SAR, 15 May 1959.

20  RSA, UJL, TEBA Collection, File 33A, Secretary, Transvaal Chamber of Mines to District Manager, WNLA, Lourenço Marques, 14 February 1930.

21  SANA, Pretoria, NTS File 201, Report of the WNLA Enquiry Commission, 1911 (Confidential), pp 3–4.

22  RSA, UJL, TEBA Collection, File 75/5, BL Thurber to Compound Manager, WNLA, Johannesburg, 3 March 1918.

23  As late as 1927–1937 there were never fewer than 70 000 black miners directly contracted to the WNLA from southern Mozambique each year; see J Penvenne, *African Workers and Colonial Racism* (Johannesburg, 1994), p 96; and for a table covering the entire period 1902–1961, see A Isaacman and B Isaacman, *Mozambique: From Colonialism to Revolution, 1900–1981* (Aldershot, 1983), p 33.

24  See RSA, UJL, TEBA Collection, File 37/1, Acting Secretary, WNLA to General Manager, SAR, 10 October 1919; and File 33A/7, Secretary, Chamber of Mines to Compound Manager, WNLA, Witbank, 7 April 1924.

## CHAPTER FIVE

1  Conventional coaches with dual entrances at both ends of the corridor slowed down the processes of entraining and detraining and therefore earned some of the ire of certain WNLA managers, who preferred the old, single-compartment, arrangement; see, for example, Republic of South Africa (RSA), University of Johannesburg Library, TEBA Collection, Secretary, Chamber of Mines to District Manager, WNLA, Lourenço Marques, 14 February 1930.

2  AH Jeeves, *Migrant Labour in South Africa's Mining Economy: The Struggle for the Gold Mines' Labour Supply, 1890–1920* (Johannesburg, 1985), p 19 [hereafter Jeeves, *Migrant Labour*].

3  RSA, UJL, TEBA Collection, File 75/5, 'Translation of Portuguese Native Contract', 1918.

4  RM Packard, *White Plague, Black Labor: Tuberculosis and the Political Economy of Health and Disease in South Africa* (Los Angeles, 1989), p 177 [hereafter Packard, *White Plague, Black Labor*].

5  Packard, *White Plague, Black Labor*, pp 69–70.

6  P Harries, *Work, Culture, Identity and Migrant Labourers in Mozambique and South Africa, c 1860–1910* (Johannesburg, 1994), pp 196–197 [hereafter Harries, *Work, Culture, Identity*].

7  An interview with Mauricio Nkome Jnr, as recorded in R First, *Black Gold: The Mozambican Miner, Proletarian and Peasant* (New York, 1983), pp 101–102.

8  See GH Pirie, 'Railways and Labour Migration to the Rand Mines: Constraints and Significance', *Journal of Southern African Studies*, Vol 19, No 4 (December 1993), pp 723–725 [hereafter Pirie, 'Labour Migration'].

9  See RSA, South African National Archives (SANA), Pretoria, SNA File 294, Acting Secretary, WNLA to Director of Customs, Pretoria, 28 October 1905 – 'natives come to our depots in Portuguese East Africa as a rule without clothing'.

10  See L Phillips to RW Schumacher, 24 February 1911, in M Fraser and A Jeeves (eds), *All that Glittered: Selected Correspondence of Lionel Phillips, 1890–1924* (Cape Town, 1977), p 237.

11  See Pirie, 'Labour Migration', p 725; and SANA, Pretoria, GNLB Vol 183, Secretary, WNLA, Johannesburg to the Director of Native Labour, 13 May 1914.

12  RSA, SANA, Pretoria, GNLB, Vol 183, General Manager, WNLA to General Manager, SAR, 8 May 1914; and Harries, *Work, Culture, Identity*, p 179.

13  See M Newitt, *A Short History of Mozambique* (Johannesburg, 2018), p 125.

14  See Arquivo Histórico de Moçambique, Maputo, File CX 140, *Fiscalisacao de Emigracão em Ressano Garcia*, 31 March 1921.

15  On fingerprints and early Mozambican desertion rates, see Jeeves, *Migrant Labour*, pp 48, 167 and 176. As late as 1959, the average desertion rate from the mines by WNLA recruits was still 2.75 per cent; see RSA, UJL, TEBA Collection, File 33A/5, General Manager, Johannesburg to Curator of Portuguese Natives, 2 October 1959.

16  See WNLA, *Report of the Board of Management for the Year ended 31st December, 1937* – Annexure 'A' – 'Analysis of "Wastage" for the Years 1902 to 1937 inclusive'.

17  RSA, UJL, TEBA Collection, Pad 1, Deserters, 19 December 1945 to January 1951; writer's signature indecipherable, letter to W Gemmill, Salisbury, dated 8 November 1949.

18  RSA, UJL, TEBA Collection, Pad 1, Deserters, 19 December 1945 to January 1951.

19  See Jeeves, *Migrant Labour*, pp 15 and 215; and SE Katzenellenbogen, *South Africa and Southern Mozambique: Labour, Railways and Trade in the Making of a Relationship* (Manchester, 1982), p 153.

20  RSA, UJL, TEBA Collection, WNLA File 132/1, BL Thurber to Secretary, WNLA, Johannesburg, 1 February 1913.

21  See especially RSA, UJL, TEBA Collection, File 33A, Secretary, Chamber of Mines to General Manager, SAR, 14 December 1926; General Manager, SAR to Secretary, WNLA, Johannesburg, 23 December 1926; Assistant Secretary, WNLA to Compound Manager, Johannesburg, 28 December 1926; and Conductor R Thomson to Compound Manager, WNLA, Johannesburg, 31 December 1926.

22  On locked trains as part of the ongoing tension that characterised the private-public problem that the WNLA transport system presented, see RSA, UJL, TEBA Collection,

Unsigned WNLA letter to Acting General Manager SAR, Johannesburg, 29 June
1910; General Manager, SAR to Secretary, WNLA, Johannesburg, 22 December 1911;
Conductors BL Thurber and R Thomson to Chief Compound Manager, WNLA,
Johannesburg, undated, 1921.

23 This paragraph is based almost entirely on J Crush, A Jeeves and D Yudelman, *South
Africa's Labor Empire: A History of Black Migrancy to the Gold Mines* (Cape Town, 1991),
p 51 (emphasis added).

24 See GH Pirie, 'Brutish Bombelas: Trains for Migrant Gold Miners in South Africa,
c 1900–1925', *The Journal of Transport History*, No 18 (1997), p 34 [hereafter Pirie, 'Brutish
Bombelas'].

25 See RSA, UJL, TEBA Collection, File 33A/5, District Superintendent, Native Recruiting
Corporation (representing WNLA) to General Manager, WNLA, Johannesburg, 31 July
1953.

26 RSA, UJL, TEBA Collection, File 33A/3, R Thomson to Compound Manager, WNLA,
Johannesburg, 3 October 1927; and Compound Manager, Johannesburg to General
Manager, WNLA, 11 October 1927.

27 RSA, UJL, TEBA Collection, File 33A/3, Secretary, Transvaal Chamber of Mines to
District Manager, WNLA, Lourenço Marques, 14 October 1927.

28 RSA, UJL, TEBA Collection, File 33A/3, Compound Manager, WNLA, Johannesburg to
General Manager, WNLA, 11 January 1929.

29 RSA, UJL, TEBA Collection, File 33A/3, Secretary, Chamber of Mines to General
Manager, SAR, 9 February 1926. '*Sic*' emphasis added by the author.

30 RSA, UJL, TEBA Collection, File 33A/3, Compound Manager, Johannesburg to General
Manager, WNLA, 11 January 1929.

31 RSA, UJL, TEBA Collection, File 33A/5, District Superintendent, Native Recruiting
Corporation to General Manager, WNLA, 31 July 1953.

32 Paragraph based on material culled from RSA, UJL, TEBA Collection, File 75/5, Manager and
Secretary, WNLA, Johannesburg to General Manager, SAR, 22 October 1918; and Assistant
Secretary, WNLA, Johannesburg to Acting General Manager, SAR, 12 December 1918.

33 See Pirie, 'Brutish Bombelas', p 33.

34 See RSA, UJL, TEBA Collection, File 33A, BL Thurber to Chief Compound Manager,
Johannesburg, 28 January 1922; File 33A/4, BL Thurber to Station Master, Booysens,
1 February 1926; and File 33A/3, Secretary, Chamber of Mines to General Manager, SAR,
9 February 1926.

35 As quoted in Pirie, 'Brutish Bombelas', p 33.

36 See RSA, UJL, TEBA Collection, File 33/A, R Thomson and BL Thurber to L du Plessis,
Acting Chief Compound Manager, WNLA, Johannesburg, 15 June 1923.

37 On some of the attendant dangers, such as having to leave the train and cross adjacent
tracks being used for shunting operations in order to get to toilets on station platforms,

see, for example, RSA, UJL, TEBA Collection, Secretary, Chamber of Mines, Johannesburg to General Manager, SAR, 18 February 1941.

38 RSA, UJL, TEBA Collection, File 33A, BL Thurber to Acting Secretary, WNLA, Johannesburg, 20 June 1926.

39 RSA, UJL, TEBA Collection, File 33A, R Thomson to General Manager, WNLA, Johannesburg, 22 May 1930.

40 RSA, UJL, TEBA Collection, File 33A, Extract from a report by R Thomson dated 24 March 1932.

41 RSA, UJL, TEBA Collection, File 33A, Manager and Secretary, WNLA, Johannesburg to General Manager, SAR, 29 June 1920.

42 RSA, UJL, TEBA Collection, File 33A, Manager and Secretary, WNLA, Johannesburg to General Manager, SAR, 3 July 1923.

43 RSA, UJL, TEBA Collection, File 33/A, R Thomson to Compound Manager, WNLA, Johannesburg, 9 June 1941.

44 RSA, UJL, TEBA Collection, File 33A/4, Secretary, Chamber of Mines to General Manager, 19 July 1993.

45 See also RSA, UJL, TEBA Collection, File 33A/5, Secretary, Chamber of Mines to General Manager, SAR, 10 July 1933; Secretary, Chamber of Mines to General Manager, SAR, 28 May 1935; Secretary, Chamber of Mines to General Manager, SAR, 26 June 1935; and General Manager, SAR to Secretary, Chamber of Mines, Johannesburg, 14 August 1936.

46 HA Junod, *The Life of a South African Tribe* (New York, 1962), pp 80 and 93 [hereafter Junod, *The Life*].

47 Harries, *Work, Culture, Identity*, p 158.

48 S Gigliotti, *The Train Journey: Transit, Captivity and Witnessing in the Holocaust* (Oxford, 2010), p 135. See also SA Ashley, 'Railway Brain: The Body's Revenge Against Progress', *Proceedings of the Western Society for French History*, Vol 31 (2003), pp 177–196.

49 A Alben, 'Eighteen Seconds to Impact', *Times Literary Supplement*, 27 March 2019.

50 See especially 'Railroad Space and Railroad Time' in Wolfgang Schivelbusch's classic *The Railway Journey: The Industrialization of Time and Space in the 19th Century* (Berkeley, 1986).

51 See, for example, RSA, NASA, Pretoria, NTS, File 201, 'Report of the WNLA Enquiry Commission', 1911, p 3; and Harries, *Work, Culture, Identity*, pp 55–58 and 152.

52 See Junod, *The Life*, p 76.

# CHAPTER SIX

1 See, among others, J McCulloch, 'Mine Medicine, Knowledge and Power on South Africa's Gold Mines', *Labor History*, Vol 54, No 4, (2013), pp 421–435, or, by the same author, at greater length but with a narrower focus, *South Africa's Gold Mines and the Politics of Silicosis* (Johannesburg, 2012) [hereafter McCulloch, 'Mine Medicine'].

2  Republic of South Africa (RSA), University of Johannesburg Library (UJL), TEBA Collection, File 10, Manager and Secretary, WNLA, Johannesburg to District Manager, Lourenço Marques, 1 August 1922.

3  From this perspective see, for example, L Weiss, 'Exceptional Space: Concentration Camps and Labor Compounds in Late Nineteenth-Century South Africa', in A Myers and G Moshenska (eds), *Archaeologies of Internment* (New York, 2011), pp 21–32; and on the first two decades of the 20th century, AH Jeeves, *Migrant Labour in South Africa's Mining Economy: The Struggle for the Gold Mines' Labour Supply, 1890–1920* (Johannesburg, 1985), p 21 [hereafter Jeeves, *Migrant Labour*].

4  N Mandela, *Long Walk to Freedom: The Autobiography of Nelson Mandela* (London, 1994), p 60 [hereafter Mandela, *Long Walk to Freedom*].

5  Mandela, *Long Walk to Freedom*, p 96.

6  See B Ngqulunga, *The Man who Founded the ANC: A Biography of Pixley ka Isaka Seme* (Cape Town, 2017), pp 58–59.

7  See RM Packard, *White Plague, Black Labor: Tuberculosis and the Political Economy of Health and Disease in South Africa* (Los Angeles, 1989), p 165 [hereafter Packard, *White Plague, Black Labor*]; and F Wilson, *Labour in the South African Gold Mines, 1911–1969* (Cambridge, 1972), p 58.

8  See, for example, P Harries, *Work, Culture, Identity and Migrant Labourers in Mozambique and South Africa, c 1860–1910* (Johannesburg, 1994), p 158 [hereafter Harries, *Work, Culture, Identity*].

9  Harries, *Work, Culture, Identity*, p 208.

10  In the UK and the USA it was only after the Second World War that these links began to be explored, and then only in the most tentative fashion possible. See, for example, LW Field, RT Ewing and DM Wayne, 'Observations on the Relation of Social of Psychosocial Factors to Psychiatric Illness among Coal-miners', *International Journal of Social Psychiatry*, Vol 3 (September 1957), pp 133–145; F Post, 'A Study of Psychiatric Illness in Coal Miners', *Journal of Mental Science*, Vol 92, No 388 (July 1946), pp 74–584; or C Wiesel and M Arny, 'Psychiatric Study of Coal Mines in the Eastern Kentucky Area', *American Journal of Psychiatry*, Vol 108 (February 1952), pp 617–624. I am indebted to Professor Robert Kaplan for drawing these sources to my attention. There appear to be no comparable sources for South Africa.

11  Paragraph based on information drawn from Harries, *Work, Culture, Identity*, pp 102–103; J Penvenne, *African Workers and Colonial Racism: Mozambican Strategies and Struggles in Lourenço Marques, 1877–1962* (Johannesburg, 1994) pp 40–41; and more especially, GR Pirio, 'Commerce, Industry and Empire: The Making of Modern Portuguese Colonialism in Angola and Mozambique, 1890–1914', unpublished PhD thesis, Department of History, University of California, Los Angeles, 1982, pp 178–194 [hereafter Pirio, 'Commerce, Industry and Empire'].

12  See, especially, 'Randlords and Rotgut, 1886–1914; The Role of Alcohol in the

Development of European Imperialism and Southern African Capitalism with Special Reference to Black Mineworkers in the Transvaal Republic', in C van Onselen, *New Babylon, New Nineveh: Everyday Life on the Witwatersrand, 1886–1914* (Johannesburg, 2011), pp 47–108 [hereafter van Onselen, 'Randlords and Rotgut'].

13  See van Onselen, 'Randlords and Rotgut', pp 72–73; and Pirio, 'Commerce, Industry and Empire', p 185.

14  Paragraph based on information drawn from Harries, *Work, Culture, Identity*, p 184; Jeeves, *Migrant Labour*, p 25; and SE Katzenellenbogen, *South Africa and Southern Mozambique: Labour, Railways and Trade in the Making of a Relationship* (Manchester, 1982), p 108.

15  See RSA, UJL, TEBA Collection, File 33A, General Manager, SAR to Secretary, WNLA, Johannesburg, 30 December 1936.

16  RSA, South Africa National Archives (SANA), Pretoria, NA Vol 193, Ref F473, E MacDonnell, British Consulate General, Lourenço Marques to Secretary of State, Foreign Affairs, London, 21 September 1915.

17  For a useful survey of the subject, see RG Forman, 'Randy on the Rand: Portuguese African Labor and the Discourse on "Unnatural Vice" in the Transvaal from the Early Twentieth Century', *Journal of the History of Sexuality*, Vol 11, No 4 (October 2002), pp 570–609.

18  In *Work, Culture, Identity*, pp 200–208, Harries offers a richly nuanced account of the strengths and limitations of same-sex mine marriages – *bukhontxana*.

19  R First, *Black Gold: The Mozambican Miner, Proletarian and Peasant* (New York, 1983), p xv.

20  See Harries, *Work, Culture, Identity*, pp 202 and 205.

21  These rough approximations are all drawn from N Ndlovu, J Murray and S Seopela, 'Damaged Goods: Return to Sender. A Review of the Historical Medical Record of Repatriated Chinese Miners', *Adler Museum Bulletin*, Vol 2, No 32 (2006), pp 18–25 [hereafter Ndlovu et al, 'Damaged Goods'].

22  See C van Onselen, *Chibaro: African Mine Labour in Southern Rhodesia, 1900–1933* (London, 1976), p 153.

23  RSA, UJL, TEBA Collection, File 30 J-1, 'Hospital Superintendents' – Circular Letter 310/40, 'Improvements in the Medical Organisation of the Witwatersrand Mines', 10 October 1940, p 2 – 'Qualified Male Mental Nurses'.

24  RSA, UJL, TEBA Collection, File 20C, Pad 1, E Pilcher, District Manager, Lourenço Marques to General Manager, WNLA, Johannesburg, 13 November 1950.

25  Interview with Alberto Fabiao Mugabe, former miner, by Tozio Mugabe, Maputo, 2 October 2018.

26  J Comaroff and J Comaroff, 'The Madman and the Migrant: Work and Labor in the Historical Consciousness of a South African People', *American Ethnologist*, Vol 14, No 2 (May 1987), pp 191. I am indebted to Professor Isak Niehaus who first brought this extraordinary illustration to my attention.

27 Taken from Ndlovu et al, 'Damaged Goods', p 19.

28 See, among many others, E Katz, *The White Death: Silicosis on the Witwatersrand Gold Mines, 1886–1910* (Johannesburg, 1994); J McCulloch, *South Africa's Gold Mines and the Politics of Silicosis* (Johannesburg, 2013); McCulloch, 'Mine Medicine', pp 421–435; and Packard, *White Plague, Black Labor.*

29 Packard, *White Plague, Black Labor*, pp 88–89.

30 Packard, *White Plague, Black Labor*, p 79.

31 Packard, *White Plague, Black Labor*, p xviii; and McCulloch, 'Mine Medicine', p 425.

32 The earliest initiative to separate miners suffering from tuberculosis from other patients appears to date back to mid-1914; see RSA, UJL, TEBA Collection, File 173/1 'Segregations of Tuberculosis Natives whilst travelling in trains'. From that source, however, it would appear that such segregation was determined to be impractical, and it may only have been at a later date that the first special coaches catering for infected patients were introduced on the Eastern Main Line. On the spread of tuberculosis into rural Mozambique after the First World War, see Packard, *White Plague, Black Labor*, pp 98–99.

33 Packard, *White Plague, Black Labor*, p xix.

34 Jeeves, *Migrant Labour*, p 235.

35 See Harries, *Work, Culture, Identity*, pp 189 and 221.

36 As quoted in T Maloka, 'Basotho and the Experience of Death, Dying and Mourning in the South African Mine Compounds, 1890–1940', *Cahiers d'Études Africaines*, Vol 38, No 149 (1998), p 25.

37 *Report of the Native Grievances Inquiry, 1913–1914* [UG 37-14], Para 245, p 33 [hereafter *Native Grievances*].

38 *Native Grievances*, paragraphs 246 and 247, at pp 33 and 34.

39 See, for example, RJ Williams, 'Creating a Healthy Colonial State in Mozambique, 1885–1915', unpublished PhD thesis, Department of History, University of Chicago, 2013, and, more especially, pp 132–149.

40 On the importance of a proper burial see, for example, HA Junod, *The Life of a South African Tribe* (New York, 1962), pp 165–169.

# CHAPTER SEVEN

1 See Republic of South Africa (RSA), University of Johannesburg Library (UJL), TEBA Collection, File 33A/3, BL Thurber to Secretary, WNLA, Johannesburg, 1 December 1924.

2 For examples spanning the period 1924–1933 see RSA, UJL, TEBA Collection, File 33A/3, BL Thurber to General Manager, WNLA, Johannesburg, 5 April 1924; File 33A/4, BL Thurber to Station Master, Booysens, 1 February 1926; File 33A/3, Compound Manager to General Manager, WNLA, Johannesburg, 11 January 1929.

3 RSA, UJL, TEBA Collection, File 33A, BL Thurber to Compound Manager, WNLA, Johannesburg, 16 May 1927.

4 Before the First World War, at a time when the separation of TB sufferers from other patients was being mooted, it was reported that 'all sick natives travel together on Thursdays'; see RSA, UJL, TEBA Collection, General Manager, SAR to Secretary, WNLA, Johannesburg, 12 June 1914; and Secretary, WNLA to General Manager, SAR, Johannesburg, 15 June 1914.

5 On 'rejects' see, for example, RSA, UJL, TEBA Collection, File 35, 'Repatriations and Rejects: East Coast Natives, March 1925–August 1957; and on 'lying down' cases, see RSA, NASA, Pretoria, GNLB Vol 178, Chief Medical Officer, WNLA to Pass Officer, WNLA, Johannesburg, 21 July 1926.

6 RSA, UJL, TEBA Collection, File 33A, R Thomson to Compound Manager, WNLA, Johannesburg, 31 December 1926.

7 RSA, UJL, TEBA Collection, File 33A/4, Manager, Mozambique to General Manager, Johannesburg, 14 February 1963.

8 RSA, UJL, TEBA Collection, File 35, District Manager, WNLA, Lourenço Marques to General Manager, WNLA, Johannesburg, 22 April 1952.

9 RSA, UJL, TEBA Collection, File 33A/4, General Manager, SAR to Secretary, WNLA, Johannesburg, 26 February 1923.

10 See RSA, UJL, TEBA Collection, File 33A/4, General Manager, SAR to General Manager, Chamber of Mines, Johannesburg, 22 September 1938; General Manager, SAR to General Manager, Chamber of Mines, 7 June 1939; and Acting Compound Manager, WNLA to Manager, WNLA, Johannesburg, 9 October 1940.

11 RSA, UJL, TEBA Collection, File 33A/4, Secretary, Transvaal Chamber of Mines to General Manager, SAR, 21 April 1951; and Chief Medical Officer, WNLA to General Manager, WNLA, 12 January 1963.

12 This paragraph based on information drawn from RSA, UJL, TEBA Collection, File 33A, BL Thurber to Compound Manager, WNLA, Johannesburg, 16 May 1927; Secretary, Chamber of Mines to General Manager, SAR, 10 July 1933; and Secretary, Chamber of Mines to General Manager, SAR, 20 November 1933.

13 See J McCulloch, 'Mine Medicine: Knowledge and Power on South Africa's Gold Mines', *Labor History*, Vol 54, No 4 (2013), p 425.

14 RSA, UJL, TEBA Collection, File 33/A, BL Thurber to General Manager, SAR, 10 December 1920 (emphases added).

15 RSA, UJL, TEBA Collection, File 33/A, BL Thurber to Manager and Secretary, WNLA, Johannesburg, 13 September 1920.

16 RSA, UJL, TEBA Collection, File 20 L, AI Girdwood, Chief Medical Officer to General Manager, WNLA, Johannesburg, 17 February 1925.

17 RSA, South African National Archives (SANA), Pretoria, GNLB Vol 178, Inspector W Walker to Director, Native Labour, 25 January 1918.

18 RSA, UJL, TEBA Collection, File 33A/3, Secretary, Transvaal Chamber of Mines to General Manager, SAR, Johannesburg, 15 August 1927.

19  See, for example, RSA, UJL, TEBA Collection, File 33A, BL Thurber to Compound Manager, WNLA, 13 July 1936.

20  RSA, UJL, TEBA Collection, File 75/2, AS 69, Acting Secretary, WNLA, Johannesburg to Compound Manager, WNLA, Witbank, 7 November 1921.

21  See RSA, UJL, TEBA Collection, File 75/5, Manager and Secretary, WNLA to General Manager, SAR, 13 March 1918; and BL Thurber to Chief Compound Manager, WNLA, Johannesburg, 15 April 1918; File 33A, BL Thurber to Chief Compound Manager, WNLA, Johannesburg, 28 January 1922; and Secretary, Transvaal Chamber of Mines to Compound Manager, WNLA, Johannesburg, 10 February 1959.

22  The most complete of the surviving sources for tallying deaths on the down-train are to be found in RSA, NASA, Pretoria, GNLB Vols 178–180. Scattered references to other 307 deaths are to be found elsewhere in RSA, UJL, TEBA Collection.

23  For a selection of these cases, see, for example, items housed in RSA, SANA, Correspondence of the Director of Native Labour, including Part 763/14/104, Sgt JTH, Komatipoort to Department of Native Affairs, 9 September 1919; Part File 763/14/104, indecipherable to Director Native Labour, 8 February 1920; Part File 763/14/104, W Walker to Director Native Labour, 25 January 1918; Part File, 762/14/104, Acting Pass Officer, Johannesburg, 23 July 1926; TAB GNLB 179/763/14/1, Director of Native Labour to Inspector, WNLA, Johannesburg, 11 March 1927; TAB GNLB 179/763/14/1, Director of Native Labour to Inspector, WNLA, Johannesburg, 30 April 1919; TAB GNLB 179/763/14/1 Part File 763/14/D104, Manager and Secretary, WNLA to Inspector, Central Department of Native Affairs, 24 October 1922.

24  RSA, UJL, TEBA Collection, Pad 1, 'Deaths of Natives en route to and from their Homes (not accidents)', Assistant Secretary, Chamber of Mines to Compound Manager, WNLA, Johannesburg, 5 February 1925.

25  See RSA, UJL, TEBA Collection, Pad 1, 'Deaths of Natives en route to and from their Homes (not accidents)', Chamber of Mines to Portuguese Curator, 16 February 1927.

26  All these can be traced in RSA, UJL, TEBA Collection, File 17/2, 'Deaths of Natives, Ressano Garcia and Lourenço Marques'.

27  RSA, UJL, TEBA Collection, File 33A, BL Thurber to Chief Compound Manager, WNLA, Johannesburg, 28 January 1922.

28  RSA, UJL, TEBA Collection, Pad 1, 'Deaths of Natives en route to and from their Homes (not accidents)'; BL Thurber to Compound Manager, WNLA, Johannesburg, 14 January 1925.

29  For examples covering the period 1919–1927, see RSA, SANA, Pretoria, GNLB Vol 178, Director of Native Labour to The Inspector, Johannesburg Central, 22 December 1919; GNLB Vol 179, unsigned note 1923; and GNLB Vol 179, Chief Inspector to Director of Native Affairs, 7 March 1927.

30  RSA, NASA, Pretoria, GNLB Vol 178, Inspector to the Director of Native Labour, Johannesburg, 11 March 1923.

31  In those cases in the 1920s, when the system was improving and where patients had expired on the 307, WNLA doctors took shelter behind professional opinions that were never subjected to further questioning, nor subjected to an independent second opinion by the state. 'The natives appeared perfectly fit to travel,' the 'native was fit to travel at the time of his repatriation', 'from his condition there was no reason to suspect that he would die *en route*', 'during his detention here he was running a normal temperature, and as this was a good omen, he was allowed to leave for home', and 'I was of the opinion he was in a fit state to travel' were all considered to offer sufficiently convincing explanations for deaths for state officials not to push any further. These quotations are taken, in order presented, from RSA, SANA, Pretoria, GNLB Vol 178, Chief Medical Officer to Pass Officer, Johannesburg, 22 July 1926; GNLB Vol 178, Director of Native Affairs Department, Johannesburg Central to the Director Native Labour, 13 December 1917; and GNLB Vol 178, AP Watkins to Acting Director of Native Affairs, Johannesburg, 13 December 1917.

32  R Ehrlich, 'The Body as History: On Looking at the Lungs of Miners', Inaugural Lecture at the School of Public Health and Family Medicine, Faculty of Health Sciences, University of Cape Town, 26 September 2007.

33  RSA, UJL, WNLA, File 75/6, District Manager, Lourenço Marques to Manager and Secretary, WNLA, Head Office, Johannesburg, 'Confidential', 2 April 1919. That letter needs to be read as part of a response to another 'Confidential' letter, from Dr AP Watkins, WNLA to W Gemmill, Manager and Secretary, WNLA, Johannesburg, 26 March 1919.

34  See, for example, the case of JA Wepner as recorded in RSA, UJL, TEBA Collection, File 30 J-1, Chief Medical Officer to Manager, WNLA, Johannesburg, 23 April 1942.

35  RSA, UJL, TEBA Collection, File 75/5, BL Thurber and R Thomson to Compound Manager, WNLA, Johannesburg, 3 March 1918.

36  RSA, UJL, TEBA Collection, File 33A, Secretary, Chamber of Mines to General Manager, SAR, Johannesburg, 20 November 1933.

37  RSA, UJL, TEBA Collection, File 33A/4, JA Gemmill to the Manager Mozambique, WNLA, 12 February 1963.

38  RSA, UJL, TEBA Collection, File 33A/4, JA Gemmill to the Manager Mozambique, WNLA, 12 February 1963.

# CHAPTER EIGHT

1  L Hyde, *Trickster Makes This World: Mischief, Myth and Art* (Edinburgh, 1998), p 253 (emphasis added).

2  Republic of South Africa (RSA), University of Johannesburg Library (UJLS), TEBA Collection, File 33A/3, Secretary, Transvaal Chamber of Mines to General Manager, SAR, Johannesburg, 26 January 1927.

3  RSA, UJL, TEBA Collection, File 178/1, Breyner & Wirth Ltd, Lourenço Marques, 'confidential' to JA Gemmill WNLA, Johannesburg, 14 May 1923.

4  See, for example, RSA, UJL, TEBA Collection, File 33A, P Gaivao, Breyner & Wirth Ltd, Lourenço Marques to General Manager, WNLA, Johannesburg, 26 March 1930.

5  See, for example, RSA, UJL, TEBA Collection, Secretary, Chamber of Mines to General Manager, SAR, 26 January 1927.

6  RSA, UJL, TEBA Collection, File 33A/7, Secretary, Transvaal Chamber of Mines to General Manager, SAR, Johannesburg, 2 May 1925.

7  RSA, UJL, TEBA Collection, File 33A/12, Assistant Secretary, WNLA to Compound Manager, Johannesburg, 5 October 1932.

8  RSA, UJL, TEBA Collection, File 37/1, R Thomson to Chief Compound Manager, WNLA, Johannesburg, 2 February 1922.

9  See RSA, UJL, TEBA Collection, File 33A/12, Secretary, Chamber of Mines to Station Master, Germiston, 30 January 1928; and File 33A/4, R Thomson to Compound Manager, WNLA, Johannesburg, 3 October 1929.

10  RSA, UJL, TEBA Collection, File 33A/5, A Nosworthy to Compound Manager, WNLA, Johannesburg, 21 February 1956

11  This paragraph is constructed from material contained in RSA, UJL, TEBA Collection, where File 33A/7 constitutes a rich vein of material on 'Thefts or Other Losses of Native Belongings' in the 1920s.

12  RSA, UJL, TEBA Collection, File 33A/7, General Manager, SAR to Secretary, WNLA, Johannesburg, 5 July 1925.

13  RSA, UJL, TEBA Collection, File 33A/7, Secretary, Chamber of Mines to the Portuguese Curator, Johannesburg, 24 July 1925.

14  See, for example, RSA, UJL, TEBA Collection, File 33A/7, Secretary, Chamber of Mines to General Manager, WNLA, Salisbury, 29 October 1956.

15  See, for example, RSA, UJL, TEBA Collection File 33A, SAR System Manager to Secretary, WNLA, Johannesburg, 9 December 1935 – '[R]epresentations have been made to this office [SAR] for the time allowed at Nelspruit Station to be extended, to allow the lessee of the refreshment stall sufficient time to cater for the natives'.

16  'As regards boys going home to Portuguese territory the Association only supplies bread, it being left to the natives themselves to buy what other form of foodstuffs they may require' – RSA, UJL, File 75/2 AS 69, Manager and Secretary, WNLA, Johannesburg to J McNaught Esq, Kaalplaats, 15 December 1922.

17  The wider context of these events, spread over several years, can be traced in RSA, UJL, TEBA Collection, File 33A/11, which is devoted to 'interference with natives' on 'the railway line'.

18  See RSA, UJL, TEBA Collection, File 37/1, R Thomson to Chief Compound Manager, Johannesburg, 18 April 1921; and File 33/8, BL Thurber to Compound Manager, WNLA,

Johannesburg, 13 July 1936.

19  RSA, UJL, TEBA Collection, File 37/1, BL Thurber (and R Thomson) to Chief
    Compound Manager, WNLA, Johannesburg, undated letter, 1921.

20  Paragraph constructed from material in RSA, UJL, TEBA Collection, File 37/1, and,
    more especially in the following letters: BL Thurber to Chief Compound Manager,
    WNLA, Johannesburg, 29 July 1918; BL Thurber to Acting Secretary, WNLA,
    Johannesburg, 29 October 1919; and File 33A/7, A Herschell to General Manager,
    WNLA, Johannesburg, 5 March 1932.

21  RSA, UJL, TEBA Collection, File 33/A, ER Carney, System Manager, SAR to Secretary,
    WNLA, Johannesburg, 9 December 1935.

22  RSA, UJL, TEBA Collection, File 33/A, BL Thurber to W Gemmill, Manager and
    Secretary, WNLA, Johannesburg, 13 September 1920 (emphasis as in the original).

23  RSA, UJL, TEBA Collection, File 33/A, District Manager, Lourenço Marques to General
    Manager, WNLA, Johannesburg, 13 February 1932, and District Manager, Lourenço
    Marques to General Manager, WNLA, Johannesburg, 20 January 1930.

24  For Mozambican deaths from silicosis during the first decades see, for example, E Katz,
    *The White Death: Silicosis on the Witwatersrand Gold Mines, 1886–1910* (Johannesburg,
    1994), pp 190–199; or RM Packard, *White Plague, Black Labor: Tuberculosis and the
    Political Economy of Health and Disease in South Africa* (Los Angeles, 1989), pp 94–95.

25  See unsigned letter to Chief Compound Manager, WNLA, Johannesburg, 31 August 1918
    (RSA, UJL, TEBA Collection, File 37/1).

26  On pickpocketing at Witbank see, eg, RSA, UJL, TEBA Collection, File 33A/5, Secretary,
    Chamber of Mines to the Post Commander, Railway Police, Witbank, 18 January 1960.

27  See 'Robbery on a Train – Native Victims', *The Star*, 5 January 1924.

28  See, for examples, RSA, UJL, TEBA Collection, File 33A/7, Secretary, Chamber of Mines
    to System Manager, SAR, Johannesburg, 4 April 1939; and Assistant Manager, WNLA to
    Curator of Portuguese Natives, Johannesburg, 29 December 1941.

29  RSA, UJL, WNLA, File 33A/7, Curator of Portuguese Natives to Manager, WNLA,
    Johannesburg, 3 December 1941.

30  RSA, UJL, TEBA Collection, File 33A/7, Secretary, Chamber of Mines to General
    Manager, SAR, Johannesburg, 30 January 1932.

31  RSA, UJL, TEBA Collection. File 75/6, Chief Compound Manager, Johannesburg to
    W Walker, Inspector, Native Affairs, 6 February 1922.

32  This narrative has been constructed from two documents to be found in RSA,
    UJL, TEBA Collection, File 33A/7, BL Thurber to Compound Manager, WNLA,
    Johannesburg, 7 January 1924; and Compound Manager, WNLA, Witbank,
    'Memorandum in Connection with theft by False Pretences from Medical Rejects on
    WNLA Train, 3rd January, 1924'.

33  See RSA, UJL, TEBA Collection, File 33A/7, Compound Manager, WNLA, Witbank
    to General Manager, WNLA, Johannesburg, 18 March 1924; and Secretary, Transvaal

Chamber of Mines to District Manager, WNLA, Lourenço Marques, 7 April 1924.

34  See also RSA, UJL, TEBA Collection, File 33A/7, Chamber of Mines to District Manager, Lourenço Marques, 9 April 1929; BL Thurber to Compound Manager, WNLA, Johannesburg, 24 April 1929; District Manager, Zoekmekaar to General Manager, WNLA, Johannesburg, 29 August 1929; and Compound Manager, Johannesburg to General Manager, WNLA, 7 February 1935.

35  RSA, UJL, TEBA Collection, File 33A/7, Compound Manager, Witbank to General Manager, WNLA, 5 April 1924.

36  The conductor concerned, BL Thurber, is the subject of a prospective biographical essay.

37  RSA, UJL, TEBA Collection, File 30/52, Manager and Secretary, WNLA, Johannesburg to BL Thurber, Esq, 20 August 1923.

38  G Clarence-Smith, *The Third Portuguese Empire, 1825–1975* (Manchester, 1985), pp 117–118.

39  SE Katzenellenbogen, *South Africa and Southern Mozambique: Labour, Railways and Trade in the Making of a Relationship* (Manchester, 1982), p 134 [hereafter Katzenellenbogen, *South Africa and Southern Mozambique*].

40  Katzenellenbogen, *South Africa and Southern Mozambique*, pp 134–135; and RSA, UJL, TEBA Collection, File 30/52, BL Thurber to General Manager, WNLA, Johannesburg, 22 August 1923.

41  See Arquivo Histórico de Moçambique, Maputo, Processa No 125, 'Troca de Dinheirios', 1922–1924.

42  As quoted in P Harries, *Work, Culture, Identity and Migrant Labourers in Mozambique* (Johannesburg, 1994), p 176.

# CHAPTER NINE

1  L Hyde, *Trickster Makes This World: Mischief, Myth and Art* (Edinburgh, 2008), p 97 [hereafter Hyde, *Trickster Makes This World*]. The other definitions of 'accident' are drawn from the *Oxford English Dictionary*.

2  See B Ngqulunga, *The Man Who Founded the ANC: A Biography of Pixley ka Isaka Seme* (Cape Town, 2017), pp 71–72.

3  Republic of South Africa (RSA), University of Johannesburg Library (UJL), TEBA Collection, File 75/5, Undated statements by BL Thurber (Conductor) and J Irving (Permanent Way Inspector) presented before the 'Enquiry held by Members of the Board of Enquiry at Waterval Boven on February 19th 1918, with reference to Accident to Native Train No 307 between Waterval Boven and Waterval Onder on the morning of 15 February, 1918' [hereafter 'Accident to Native Train No 307'].

4  'Accident to Native Train No 307', statement by Noble Dixon, Driver.

5  'Accident to Native Train No 307', p 1.

6  'Accident to Native Train No 307', statement by Noble Dixon, Driver.

7 'Accident to Native Train No 307', statement by Baron (*sic*) Lee Thurber.

8 'Accident to Native Train No 307', statement by Baron (*sic*) Lee Thurber.

9 See 'Finding', 'Accident to Native Train No 307.'

10 See 'Finding', 'Accident to Native Train No 307'.

11 RSA, UJL, TEBA Collection, File 75/5, Manager and Secretary, WNLA, Johannesburg to District Manager, WNLA, Lourenço Marques, 6 July 1918.

12 WNLA, *Report of the Board of Management for the Year ended 31st December 1949*, p 3.

13 See, especially, the comments of a survivor, the Rev Carlos Matshinye (86), as reported on by J Ingram in *The Highveld Gazette*, December 2014.

14 RSA, Johannesburg, South African Railways Heritage Centre, 'Report of Departmental Senior Officers' Enquiry Board into the Circumstances attending the Derailment of Mixed (Natives) Train No 513 in the section Waterval Boven–Ondervalle on 15th November, 1949', p 45 [hereafter 'Derailment of Mixed (Natives) Train No 513'].

15 'Derailment of Mixed (Natives) Train No 513', p 45.

16 'The Board felt the late start of the train may be of some importance for the reason that if the enginemen were responsible therefore, there might have been a tendency on their part to endeavour to recover the eight minutes on the journey to Komatipoort' – paragraph 84, 'Derailment of Mixed (Natives) Train No 513', p 26.

17 See, especially, paragraphs 115–125, pp 37–41, 'Cause of the Accident', in 'Derailment of Mixed (Natives) Train No 513'.

18 'Derailment of Mixed (Natives) Train No 513', paragraphs 12 and 151.

19 'Derailment of Mixed (Natives) Train No 513', paragraphs 15 and 153.

20 RSA, UJL, TEBA Collection, File 21A Sup, undated 'Memo for Mr JA Gemmill'.

21 'Derailment of Mixed (Natives) Train No 513', paragraph 167(b).

22 Hyde, *Trickster Makes This World*, p 100.

## CONCLUSION

1 Republic of South Africa (RSA), University of Johannesburg Library (UJL), TEBA Collection, File 199, 'Native Unrest'. See also P Alexander, 'Oscillating Migrants, "detribalised" families and militancy: Mozambicans on Witbank Collieries, 1918–1927', *Journal of Southern African Studies*, Vol 27, No 3 (2001), pp 505–525.

2 See RSA, UJL, TEBA Collection, File 199 'Native Unrest'.

3 For examples, see RSA, UJL, TEBA Collection, File 199, 'Native Unrest'.

4 RSA, UJL, TEBA Collection, File 199, 'Native Unrest', and, more especially, District Manager, Lourenço Marques, to Manager, Head Office, Johannesburg, 21 September 1918; and Assistant District Manager, Lourenço Marques, 23 September 1918.

5 See interview with Maurcio Nkome Jnr, as recorded in R First, *Black Gold: The Mozambican Miners, Proletarian and Peasant* (New York, 1983), pp 100–107.

6 Quotations taken from the commemorative booklet, *Salute to the Pioneers: Jubilee of the Opening of the Railway Line between Lourenço Marques and Pretoria, officially opened July 1895.*

7 H Kaiser, 'The Baghdad Railway and the Armenian Genocide, 1915–1916: A Case Study in German Resistance and Complicity', in RG Hovannisian (ed), *Remembrance and Denial: The Case of the Armenian Genocide* (Detroit, 1999), p 75. I am indebted to Funda Soysal for drawing this article to my attention.

8 Paragraph derived almost entirely from S Gigliotti, *The Train Journey: Transit, Captivity and Witnessing in the Holocaust* (Oxford, 2009).

9 In this broader context, and on the dangers involved in moving too easily between colonial practices and the Holocaust experience, see R Gerwarth and S Malinowski, 'Hannah Arendt's Ghosts: Reflections on the Disputable Path from Windhoek to Auschwitz', *Central European History*, Vol 42, No 2 (June 2009), pp 279–300.

10 See, especially, I Niehaus's excellent 'Witches and Zombies of the South African Lowveld: Discourse, Accusations and Subjective Reality', *The Journal of the Royal Anthropological Institute*, Vol 11, No 2 (June 2005), pp 191–210.

11 See R Shaw, 'The Production of Witchcraft/Witchcraft as Production: Memory, Modernity and the Slave Trade in Sierra Leone', *American Ethnologist*, Vol 24, No 4 (November 1997), pp 856–876 [hereafter Shaw, 'The Production of Witchcraft']. I am indebted to Professor I Niehaus for drawing this article to my attention.

12 I Niehaus with E Mohlala and K Shokane, *Witchcraft, Power and Politics: Exploring the Occult in the South African Lowveld* (London, 2001), pp 72–73.

13 M Taussig, as quoted in Shaw, 'The Production of Witchcraft', p 868.

# Select Bibliography

## ARTICLES

Allina, E, '"Fallacious Mirrors": Colonial Anxiety and Images of African Labor in Mozambique, c 1929', *History in Africa*, Vol 24 (1997), pp 9–52.

Ashley, SA, 'Railway Brain: The Body's Revenge Against Progress', *Proceedings of the Western Society for French History*, Vol 31 (2003), pp 177–196.

Comaroff, J and Comaroff, J, 'The Madman and the Migrant: Work and Labor in the Historical Consciousness of a South African People', *American Ethnologist*, Vol 14, No 2 (May 1987), pp 109–209.

Eidelberg, PG, 'The Breakdown of the 1922 Lourenço Marques Port and Railway Negotiations', *The South African Historical Journal*, Vol 8, No 1 (2000), pp 104–118.

Gerwarth, R and Malinowski, S, 'Hannah Arendt's Ghosts: Reflections on the Disputable Path from Windhoek to Auschwitz', *Central European History*, Vol 42, No 2 (June 2009), pp 279–300.

Haarhoff, J, 'Conquering the Escarpment: Railway Engineering in the Elands River Valley', *Civil Engineering* (May 2017), pp 20–30.

Harris, M, 'Labour Emigration Among the Moçambique Thonga: Cultural and Political Factors', *Africa*, Vol 29, No 1 (January 1959), pp 50–66.

Jeeves, AH, 'New Labour Frontiers for Old: Black Migrants to the South African Gold Mines, 1920–1985', *Journal of Southern African Studies*, Vol 13, No 1 (October 1986), pp 102–104.

McCulloch, J, 'Mine Medicine, Knowledge and Power on South Africa's Gold Mines', *Labor History*, Vol 54, No 4 (2013), pp 421–435.

Ndlovu, N, Murray, J and Seopela, S, 'Damaged Goods: Return to Sender. A Review of the Historical Medical Record of Repatriated Chinese Miners', *Adler Museum Bulletin*, Vol 2, No 32 (2006), pp 18–25.

Nevinson, HW, 'The New Slave Trade', *Harper's Monthly Magazine*, Vol 111, No 663 (August 1905).

Pirie, GH, 'Railways and Labour Migration to the Rand Mines: Constraints and Significance', *Journal of Southern African Studies*, Vol 19, No 4 (December 1993), pp 713–730.

Pirie, GH, 'Brutish Bombelas: Trains for Migrant Gold Miners in South Africa, c 1900–1925', *The Journal of Transport History*, No 18 (1997), pp 31–44.

Roesch, O, 'Migrant Labour and Forced Labour Production in Southern Mozambique: The Colonial Peasantry of the Lower Limpopo Valley', *Journal of Southern African Studies*, Vol 17, No 2 (June 1991), pp 230–270.

Smith, AK, 'The Idea of Mozambique and its Enemies, c 1890–1930', *Journal of Southern African Studies*, Vol 17, No 3 (September 1991).

Yudelman, D and Jeeves, A, 'New Labour Frontiers for Old: Black Migrants to the South African Gold Mines, 1920–1985', *Journal of Southern African Studies*, Vol 13, No 1 (October 1986), pp 102–104.

# BOOKS

Clarence-Smith, G, *The Third Portuguese Empire, 1825–1975* (Manchester, 1985).

Crush, J, Jeeves, A and Yudelman, D, *South Africa's Labor Empire: A History of Black Migrancy to the Gold Mines* (Cape Town, 1991).

De Klerk, WA, *The Puritans in Africa: The Story of Afrikanerdom* (London, 1975).

Duffy, J, *Portuguese Africa* (Cambridge, MA, 1959).

First, R, *Black Gold: The Mozambican Miner, Proletarian and Peasant* (New York, 1983).

Frankel, SH, *Investment and the Return to Equity Capital in the South African Gold Mining Industry, 1887–1965* (Cambridge, MA, 1967).

Gigliotti, S, *The Train Journey: Transit, Captivity and Witnessing in the Holocaust* (Oxford, 2009).

Gordon, CT, *The Growth of Boer Opposition to Kruger, 1890–1895* (London, 1970).

Gorski, PS, *The Disciplinary Revolution: Calvinism and the Rise of the State in Early Modern Europe* (London, 2003).

Harries, P, *Work, Culture, Identity and Migrant Labourers in Mozambique and South Africa, c 1860–1910* (Johannesburg, 1994).

Hyde, L, *Trickster Makes This World: Mischief, Myth and Art* (Edinburgh, 2008).

Isaacman, A and Isaacman, B, *Mozambique: From Colonialism to Revolution, 1900–1981* (Aldershot, 1983).

Jeeves, AH, *Migrant Labour in South Africa's Mining Economy: The Struggle for the Gold Mines' Labour Supply, 1890–1985* (Johannesburg, 1985).

Junod, HA, *The Life of a South African Tribe* (New York, 1962).

Katz, E, *The White Death: Silicosis on the Witwatersrand Gold Mines, 1886–1910* (Johannesburg, 1994).

Katzenellenbogen, SE, *South Africa and Southern Mozambique: Labour, Railways and Trade in the Making of a Relationship* (Manchester, 1982).

Newitt, M, *A Short History of Mozambique* (Johannesburg, 2018).

Niehaus, I with Mohlala, E and Shokane, K, *Witchcraft, Power and Politics: Exploring the Occult in the South African Lowveld* (London, 2001).

Packard, RM, *White Plague, Black Labor: Tuberculosis and the Political Economy of Health and Disease in South Africa* (Los Angeles, 1989).

Penvenne, J, *African Workers and Colonial Racism: Mozambican Strategies and Struggles in Lourenço Marques, 1877–1962* (London, 1985).

Porter, A, *Victorian Shipping, Business and Imperial Policy: Donald Currie, the Castle Line and Southern Africa* (Woodbridge, 1986).

Schivelbusch, W, *The Railway Journey: The Industrialization of Time and Space in the 19th Century* (Berkeley, 1986).

Spence, CF, *The Portuguese Colony of Moçambique* (Cape Town, 1951).

Vail, L and White, L, *Capitalism and Colonialism in Mozambique: A Study of Quelimane District* (London, 1980).

van Onselen, C, *Chibaro: African Mine Labour in Southern Rhodesia, 1900–1933* (London, 1976).

van Onselen, C, *New Babylon, New Nineveh: Everyday Life on the Witwatersrand, 1886–1914* (Johannesburg, 2001).

van Onselen, C, *Showdown at the Red Lion: The Life and Times of Jack McLoughlin, 1859–1910* (Johannesburg, 2015).

van Onselen, C, *The Cowboy Capitalist: John Hays Hammond, the American West & the Jameson Raid* (Johannesburg, 2017).

Vaughan, GE, *Portuguese East Africa* (London, 1952).

Wilson, F, *Labour in the South African Gold Mines, 1911–1969* (Cambridge, 1972).

## CHAPTERS IN BOOKS

Kaiser, H, 'The Baghdad Railway and the Armenian Genocide, 1915–1916: A Case Study in German Resistance and Complicity', in RG Hovannisian (ed), *Remembrance and Denial: The Case of the Armenian Genocide* (Detroit, 1999), pp 67–111.

Weiss, L, 'Exceptional Space: Concentration Camps and Labor Compounds in Late Nineteenth-Century South Africa', in A Myers and G Moshenska (eds), *Archaeologies of Internment* (New York, 2011), pp 21–32.

## THESES

Pirio, GR, 'Commerce, Industry and Empire: The Making of Modern Portuguese Colonialism in Angola and Mozambique, 1890–1914'. Unpublished PhD thesis, Department of History, University of California, Los Angeles, 1982.

Williams, RJ, 'Creating a Healthy Colonial State in Mozambique, 1885–1915'. Unpublished PhD thesis, Department of History, University of Chicago, 2013.

# A NOTE ON THE PRIMARY SOURCES

The primary source for this study is the records of the Witwatersrand Native Labour Association (WNLA), which are contained in The Employment Bureau of Africa (TEBA) historical papers housed in the library of the University of Johannesburg. The records are thin for the formative years of the WNLA, that is, from 1902 through to about 1910 when the coming of Union appears to have made for better recording and preservation. But, even then, the records are extremely uneven from 1910 through the mid-1920s, after which they improve somewhat in terms of both quantity and quality.

Within the TEBA records, the documentation on the train journeys is incomplete in that the reports of the train conductors appear not to have survived in consolidated format. These vital reports of the train conductors are thus incomplete, nor is there any systematic record, for example, of the number of 'sick boys' that were repatriated in the hospital coaches, nor, for that matter, of the number of patients who were insane. Such information all has to be gleaned from careful reading of other sources spread throughout the larger collection.

Wherever possible, the WNLA records have been supplemented with a collection of fragmentary records from government agencies or departments. But they too are incomplete insofar as they relate to the train, which, in truth, fell into some sort of legal no-man's land. Bits and pieces of the story of the train journey between Johannesburg and Lourenço Marques can be reconstructed from fragmentary records that come from the reports of the Director of Native Labour, the District Surgeon at Komatipoort, the South African Police and officials from the Native Affairs Department, but here, too, there is no single, dense, source of continuous primary data.

# Index

Page numbers in *italics* refer to photos, maps and illustrations.